Advanced Coating Material for Heritage Preservation

Advanced Coating Material for Heritage Preservation

Editors

Yulan Hu
Shiqiang Fang

MDPI

Basel • Beijing • Wuhan • Barcelona • Belgrade • Novi Sad • Cluj • Manchester

Editors
Yulan Hu
Zhejiang University
Hangzhou
China

Shiqiang Fang
Ningbo University of Finance & Economics
Ningbo
China

Editorial Office
MDPI
St. Alban-Anlage 66
4052 Basel, Switzerland

This is a reprint of articles from the Special Issue published online in the open access journal *Coatings* (ISSN 2079-6412) (available at: https://www.mdpi.com/journal/coatings/special_issues/advanced_coating_material_for_heritage_preservation).

For citation purposes, cite each article independently as indicated on the article page online and as indicated below:

Lastname, A.A.; Lastname, B.B. Article Title. *Journal Name* **Year**, *Volume Number*, Page Range.

ISBN 978-3-7258-0429-0 (Hbk)
ISBN 978-3-7258-0430-6 (PDF)
doi.org/10.3390/books978-3-7258-0430-6

Contents

About the Editors

Yulan Hu

Yulan Hu is an associate professor at Zhejiang University's School of Art and Archaeology. Her current research focuses on the biotechnology-based conservation of cultural heritage, which includes detecting cultural relic materials and microorganisms, preventing and controlling cultural relic microbes, and restoring and protecting cultural relics.

Shiqiang Fang

Shiqiang Fang is an associate professor in the School of Humanities, at the Ningbo University of Finance and Economics. His current research focuses on the preservation and monitoring of immovable relics, encompassing rock-carved relics, wall paintings, prehistoric architecture, inorganic cementitious materials, as well as archaeology related to science and technology.

coatings

Review

Comparison and Research Progress of Protein Detection Technology for Cultural Relic Materials

Qian Wu [1], Bingjian Zhang [1,2] and Yulan Hu [1,*]

[1] Department of Cultural Heritage and Museology, Zhejiang University, Hangzhou 310028, China; zhangbiji@zju.edu.cn (B.Z.)
[2] Department of Chemistry, Zhejiang University, Hangzhou 310027, China
* Correspondence: hu_yulan@zju.edu.cn

Abstract: The analysis of historical materials is an essential component of cultural heritage conservation. Protein was one of the most important and pervasive organic substances in ancient human societies. Through the qualitative and quantitative examination of protein-based materials, it is possible to clarify their source and functions of substances in cultural heritages, investigate the manufacturing technology of cultural heritage, and identify their deterioration mechanism. On the basis of these analyses, corresponding measurements are therefore feasible. Currently, mass spectrometry, chromatography, spectroscopy, nuclear magnetic, proteomics, and immunoassay are used to analyse protein materials. Proteomics techniques and enzyme-linked immunosorbent assay (ELISA) technology are two of the most common methods for detecting ancient proteins. This article discusses the evolution of protein component detection in ancient materials, as well as the implementation of proteomics and ELISA techniques for the analysis of proteins. In addition, the characteristics of these two techniques were contrasted in order to propose the most recent analytical techniques and the direction of future research.

Keywords: cultural heritage conservation; protein; proteomics; enzyme-linked immunosorbent assay (ELISA)

Citation: Wu, Q.; Zhang, B.; Hu, Y. Comparison and Research Progress of Protein Detection Technology for Cultural Relic Materials. *Coatings* **2023**, *13*, 1319. https://doi.org/10.3390/coatings13081319

Academic Editor: Csaba Balázsi

Received: 8 June 2023
Revised: 16 July 2023
Accepted: 20 July 2023
Published: 27 July 2023

1. Introduction

In the conservation of cultural heritage and archaeological excavations, the materials, and especially the key components, such as the binder of murals and paintings and the organic components of building mortars, are not only closely related to human production life, material life, spiritual life, and all aspects of social life, but they also serve as significant carriers of historical information. In addition, they are the focus of research on traditional building techniques, the prevention and control of the deterioration of cultural relic materials, the elucidation of the physical and chemical causes of material deterioration, and the development of future protective measures based on analysis and testing [1].

Proteins are one of the most significant components of cultural heritage artefacts and are frequently employed as essential additives [2–4]. Additionally, proteins can be used to restore and reinforce cultural heritage artefacts [5–7]. It is difficult to analyze protein-containing cultural heritage artefacts due to the limited number of available samples or the low protein content of the samples. Moreover, after prolonged exposure to multiple environmental factors, proteins encounter a variety of issues, including ageing, degradation, and contamination [8,9]. The earliest research on protein detection in artefacts of cultural heritage dates back to the 1950s [10]. Archaeologists used mass spectrometry (MS) to detect amino acids, the building blocks of proteins, in archaeological and paleontological artefacts [11]. Since the 1980s, physical chemistry techniques such as mass spectrometry, chromatography, and spectroscopy have been utilised extensively in the field of cultural heritage research and conservation. However, these methods have the disadvantage of requiring a large number of samples, making it challenging to distinguish the complex protein components in cultural heritage objects and lacking species specificity [12]. In recent years, with the development of biotechnology, bioinformatics technology, and MS, proteomics techniques and enzyme-linked immunosorbent assay (ELISA) techniques have been increasingly applied to the study of proteins in cultural relics materials, and these two techniques offer high sensitivity and low detection limits [13].

In addition to discussing the evolution of protein analysis in the field of cultural patrimony, this article will describe two cutting-edge techniques: proteomics and ELISA technology. Additionally, a general overview of the challenges and prospective directions of protein identification in cultural heritage materials will be presented.

2. The Role of Protein Analysis in the Protection of Cultural Heritage

2.1. Types of Cultural Heritage Containing Protein

Based on their composition, cultural relic materials can be classified as organic or inorganic. Both categories of ancestry share a substantial number of protein components (Table 1). For example, gum is required to enhance the writing effect on biological cultural heritage objects because the paper has numerous pores between the fibres, which easily generates a halo in the ink and affects writing and painting. The utilised adhesive is derived from either plants or animals and contains either soy protein or collagen [14–16]. The lacquer of lacquerware cultural heritage artefacts contains glycoproteins [17]. Also present in bone cultural heritage artefacts is collagen [18]. A significant number of organic cultural heritage objects are natural macromolecular proteins. Silk protein, for example, is the primary component of silk [19], whereas collagen and keratin are the primary components of leather and fur cultural heritage objects [6,18].

Table 1. Types of cultural heritage artifacts and proteins contained therein.

Cultural Heritage Material Composition	Cultural Heritage Category	Protein Containing Part	Protein Type
Organic	Paper	Glue in paper or silk	Collagen; Soybean protein
	Textile; Fur	Silk products; Wool fabric	Silk protein; Keratin
	Lacquerware; Woodware; Bambooware	Lacquer artifacts	Glycoprotein
	Leather	Leather products; Parchment	Collagen
	Carcass	Mummified corpse; Wax corpse; Wet corpse;	Collagen
	Bone; antler	Animal bones, teeth, and horns	Collagen
Inorganic	Ceramics; bricks	Pottery adhesive	Collagen; Ovalbumin; Casein
	Painted murals	Colored drawing; Mortar	Collagen; Ovalbumin; Casein

Proteins are frequently utilized as a binder in the production and restoration of inorganic cultural heritage objects, such as in the bonding of ceramic masonry artifacts, the strengthening of building materials, and the fixation of pigments in painted cultural heritage objects [2,12]. These cement-based building materials may be found easily and are commonly employed at historical and archaeological sites around the globe. Chinese architectural cement, murals, ancient architectural paintings, and pottery paintings frequently use animal glue, egg whites, and other materials as binders; these materials were later examined and found to contain collagen, ovalbumin, and other materials [2,3]. When creating dry murals and tempera paintings in the West, egg whites and milk were frequently utilized. Ovalbumin and casein have also been detected [12,20].

2.2. Characteristics of Protein-Containing Cultural Relics Materials

Physical, chemical, and biological elements in the environment can cause proteins in cultural relic materials to age and degrade. The breakdown of hydrogen and peptide bonds and the disruption of internal protein structures, which result in protein fragmentation, are induced by external stimuli or enzymatic action. For further analysis, protein fragmentation presents difficulties. Protein translational modifications (PTMs), which also cause protein fragmentation, modify the structure of proteins and alter their chemical and biological properties. The characteristics of peptides and proteins are impacted by these structural changes in significant chemical and biological ways. As a result, modifications to the side chains of some proteins can also be used as markers for the degradation of proteins as they age. One biomarker of protein degradation and age, for instance, is protein deamidation in cultural relic materials. The spontaneous nonenzymatic deamidation of glutaminyl and asparaginyl residues has been shown to alter the structure of proteins and peptides [21,22]. In addition to causing protein breakdown, these PTMs also make it challenging to distinguish between the various protein species found in cultural relic materials.

For instance, silk protein makes up the majority of silk. It is influenced by elements such as pH, microbes, oxygen, ultraviolet light, temperature, and humidity because it is a naturally occurring macromolecular protein. Silk cultural relic materials age and deteriorate as a result, losing their original form and luster and turning yellowish, brittle, or even carbonized [13,19]. This is caused by the degradation of the proteins' heavy chain and light chain, P25 chain, crystalline area, and non-crystalline region. Collagen that has been gelatinized and either been tanned or not is typically what makes up leather cultural relic materials. The acids, alkalis, and salts in the water breakdown the protein in the leather when leather cultural relic items are in a humid environment, such as when submerged in groundwater for a prolonged period of time; hence, the state is frequently quite fragile [23,24]. The morphology of leather coils, deterioration worsens, collagen and fat gradually breakdown, and leather volume diminishes, becoming thinner and more brittle over time, according to examinations of the ageing pattern of leather cultural relic materials [24].

A low temperature is ideal for preserving proteins since it slows down their aging and breakdown. In these circumstances, collagen molecules fold into a chain area, preventing protease attack and delaying protein decomposition. It has been demonstrated that collagen solutions hydrolyze at their lowest rates when the pH is between 5–7 [16]. The protein components that exist in low temperatures, have pH levels that are mildly alkaline to neutral, and are in airtight circumstances are therefore easier to preserve and analyze as residues because well-preserved collagen is also found in enclosed bone samples.

2.3. The Significance of Protein Analysis for the Protection of Cultural Relics Materials

First, protein analysis can aid in understanding the function of the protein in materials used to create cultural relics. For instance, egg white mortar, which is frequently used in construction and building decorating, can be used to track how a substance changes over time by looking for protein components in the mortar. According to research, egg whites can help to aerate, bind, sterilize, and waterproof the mortar. The interfacial activity of protein molecules and hydrated products, as well as the regulatory function of biomineralization, are related to the mechanism of these effects [2,3,7].

Second, protein analysis can show how the artifact was made. For instance, protein analysis enables the actual condition of the cultural relic materials to be compared with the records to validate the authenticity of the records. Some relics contain a preserved production formula detailing the components and production process. Vasily Yakovlevich

Struve wrote documents by hand from 1793 to 1864. In 2021, Pankin et al. [25] studied the dried film used for the paper connection and discovered that its constituents were gelatine and fish glue. This outcome is in line with the typical ingredients used in dry film making that were listed in a 19th-century encyclopedia.

In addition, protein analysis can also show how proteins age and degrade in materials from cultural relics. Collagen, for instance, is a key ingredient in leather cultural relic materials. Collagen is highly vulnerable to severe distortion, fracture, and interweaving caused by high temperatures, according to an analysis of the proteins in leather cultural relic materials. Chemical forces also break the hydrogen bonds and peptide chains of collagen, making leather cultural relic materials tougher, more fragile, and even fully ruined.

Finally, resource and approach for associated restoration and conservation activities, protein component analysis is useful. In the case of leather cultural relic materials, for instance, the degree of deterioration of the leather is assessed starting from the retention status and distribution law of the polypeptides of collagen degradation products, and a new protection method is proposed: treating degraded leather with a mixture of glutaraldehyde and tannins can strengthen the inter- and intramolecular hydrogen bond and covalently cross-linked collagen molecules to achieve reinforcement [26]. Protein component analysis can also eliminate any remaining protein clumps and offer a solution without harming the cultural heritage artifact [27]. For the first time, Paul Banks removed remaining gelatine from paper in 1969 using collagenase, a highly specialized protease [28]. Since the egg white varnish coating on the 16th-century painting was insoluble in organic solvents and mechanical removal might have irreparably ruined the original work, the enzyme removal procedure was employed while repainting the damaged area [29]. In recent years, biological purification and collagenase protease treatment have been employed successfully to remove the thick gauze bound with animal glue that was used to protect murals during World War II [30]. Different proteases may break down various kinds of proteins, so suitable proteases must be chosen in accordance with the findings of the protein analysis.

2.4. Protein Detection Method in Cultural Relics Materials

The small number of samples available for the analysis of proteins-containing cultural relic materials, the low protein content of those samples, and the difficulties in isolating proteins that are affected by aging and degradation make this approach difficult to use. Moreover, the majority of protein samples are mixtures. As a result, precise detection of protein components in mixtures is also required, in addition to efficient and sensitive approaches that can evaluate small volumes of samples [31]. Although chromatography, spectroscopy, and other technologies have been employed for protein detection, the chromatographic technology's pre-treatment process is complicated, necessitates a large number of samples, and has a high sample loss rate [32]. Furthermore, the signals produced by the spectrum in samples of cultural relic materials are subject to interference from other substances, which makes it difficult to recognize proteins, and the majority of ancient protein materials lack standard spectra, which makes it challenging to identify species [33].

Proteomics and ELISA are two recently created cutting-edge protein analysis technologies. Both approaches are very sensitive, specific, and capable of detecting proteins in complicated mixtures. Both have been widely employed [8,13,34].

3. Proteomics Technology

In 1994, Marc Wilkins, a student at Maquaire University in Australia, coined the term proteome. It refers to all proteins that are expressed by the genes of cells or tissues under specific conditions [34]. various protein expressions at various stages, locations, and environments contribute to the complexity of the proteome, making the protein group extremely complex. Proteomics is the study of the functions, structures, and interactions of all proteins expressed by genes using protein separation, mass spectrometry (MS), and biological information technologies [35,36].

Expression proteomics, which examines the qualitative and quantitative expression of proteins in cells, organs, or organisms, is the most fundamental research in proteomics [37]. Figure 1 depicts the principal processes as protein extraction and separation, MS analysis, and retrieval of biological information databases [38]. Select a suitable method for protein extraction from the sample, and then separate the extracted proteins. In the separation procedure, HPLC can be used for independent separation, or HPLC–MS can be used in series [39]. The specifics will be provided in Section 3.1.

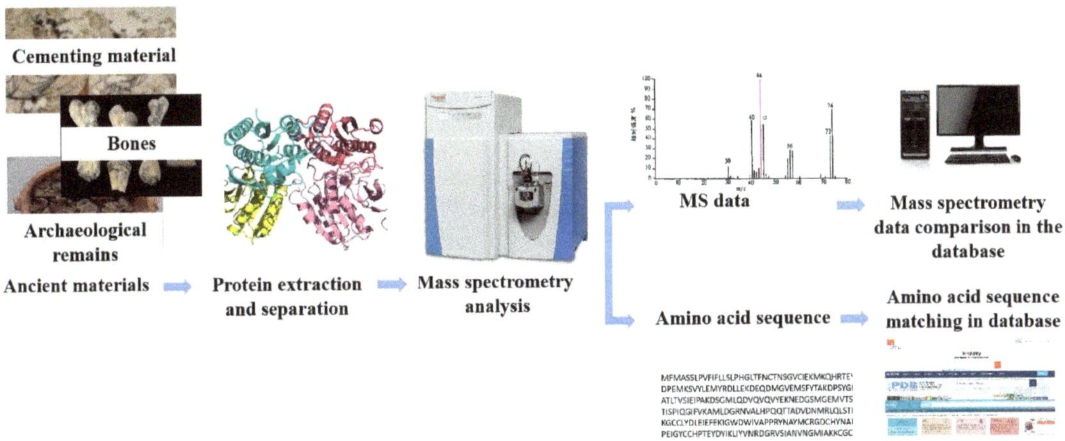

Figure 1. Main processes of proteomics.

MS analysis is the cornerstone of proteomics technology, and it consists of two distinct techniques. The top-down method analyses multiple complete proteins in a mélange, including large segment proteins, modified proteins, and peptides. The steps are as follows: First, the proteins are separated by reversed-phase liquid chromatography, and then, after ionisation and dissociation, they are analysed. This method is applicable to both sequence analysis and analysis of post-translational modifications. Sequence coverage is substantial, and sample preparation is relatively straightforward. However, there are some physical and technical challenges associated with using this method to analyse cultural relic materials. On the one hand, proteins in cultural relic materials are typically aged and degraded, and it is challenging to extract and purify complete proteins from these materials. On the other hand, mass spectrometry cannot detect aged proteins sensitively [39]. Second, the bottom-up method, also known as the shotgun method, employs a protease such as trypsin to cleave the protein backbone at specific amino acids to generate peptides, which are then analysed by mass spectrometry to derive amino acid sequence information [40]. This technique is suitable for the analysis of complex samples and has a high throughput [8]. Consequently, this technique is presently the most prevalent and mature method for protein identification, particularly in the analysis of ancient proteins. Depending on the testing requirements, various MS options can be selected or used in series due to the rapid development of instruments [38,41]. The ionisation sources of a mass spectrometer are primarily of two types: electric spray ionisation (ESI) and matrix assisted laser desorption ionisation (MALDI). These two soft ionisation techniques enable the analysis of large molecular weight molecules and proteins with post-translational modifications. Both MALDI and ESI have advantages and disadvantages and can be selected based on the sample's characteristics and the type of information needed. In addition to the ion source, the mass spectrometer consists of a mass analyzer, a detector, and a device for processing data. The mass analyzer is an essential component of the mass spectrometer, which sorts particles according to their mass-to-charge ratios (m/z). In Section 3.2, the functions and selection of the mass spectrometer will be discussed.

The detected original data can then be compared to the database to identify proteins. After digestion, each protein yields a unique set of peptides; the molecular weights of these peptides constitute the peptide fingerprint profile (PMF) of this protein. Using matrix-assisted laser desorption ionization–time-of-flight MS (MALDI-TOF-MS), for instance, we can detect the relative molecular weight of peptide segments, acquire the PMF, and then identify protein species by comparing the PMFS to a standard database. Alternatively, ESI-MS can be used to determine the amino acid sequences of various peptide segments. For protein identification, bioinformatics techniques can be used to compare the amino acid sequence analysis results with the amino acid sequences in the protein database. Using various data processing software and module analysis techniques, it is possible to detect the protein species present in a sample [42,43]. Or, tandem mass spectrometry (MS/MS or MS2), which couples two or more mass analyzers via additional reaction stages, can be utilised. Before accessing the mass spectrometer for examination, proteins must be enzymatically cleaved into peptide segments. After enzyme digestion, the mass of the peptide segment is determined by primary mass spectrometry, and significant high-abundance peptide fragments are assessed by secondary mass spectrometry. After colliding with a high flow rate of inert gas, secondary mass spectrometry dissociates peptide segments, resulting in smaller peptide fragment ions. A detector analyses the fragment ions to determine the amino acid sequence of the peptide segment. MS/MS can obtain accurate and high-quality data from abundant peptide fragment ions, enhancing the reliability of subsequent database retrieval and enabling the detection of trace proteins in mixtures.

With the gradual development of MS technology and the improvement of protein databases, proteomics technology has begun to be applied to the protein analysis of cultural relic materials; for instance, protein analysis was used to identify protein cements in paintings [44] in the early 2000s. The approach has been implemented in the field of archaeology, for example in the identification of animal remains and the analysis of animal and plant products. In proteomics research on cultural relic materials, protein extraction technology determines the efficacy of detection due to the rarity and value of the relics.

3.1. Protein Extraction

How to extract trace quantities of proteins is the first issue that must be resolved for protein identification in materials from cultural relics. Noemi Proietti used SDS-PAGE to separate and analyse the protein components in paper samples from the Camerino Fabriano region [45] in 2020, comparing the protein recovered from the samples to cellulose that had been treated with animal gelatin. The presence of a band with the same molecular weight as the standard sample indicated the presence of animal gelatin. RP-HPLC was then utilised to analyse the separated proteins. Regarding the RP-HPLC outcomes, both analysed samples exhibit chromatographic peaks with UV spectra indicative of protein material. The results indicate that the paintings contain animal gelatin. However, such procedures typically require a large number of samples, and the resulting information regarding species is frequently unclear. In an effort to reduce the weight of the sample, Caroline Tokarski [46] utilised an extraction technique that breaks down proteins from cultural relics into peptide compounds in 2006. Different extraction solutions (HCl, HCOOH, NH3, NaOH) and conditions under ultrasonic immersion, mortar grinding, and resin grinding were investigated, as were the extraction techniques for protein-cemented materials in Renaissance art paintings. The finest extraction results were obtained by grinding the grinding resin in an aqueous solution of 1% trifluoroacetic acid and performing a multi-step ultrasonic bath, according to the authors. The protein extraction efficacy was then evaluated using matrix-assisted laser desorption ionization–time-of-flight mass spectrometry (MALDI–TOF MS). Enzymatic hydrolysis of the extracted protein yields a peptide mixture, which is subsequently analysed and identified using nano-liquid chromatography (nano-LC), nano-electrospray ionisation (nano-ESI), quadrupole time-of-flight mass spectrometry (Q-q-TOF-MS), and mass spectrometry (MS). The results demonstrated that the method only required 10 g of samples to identify the binder in the painting as ovalbumin. This method employs a small sample size and does not hydrolyze the protein into amino acids during extraction, preserving more information about the genus and species.

Similarly, the extraction of proteins from cultural relics should be as minimally invasive as feasible; therefore, a microwave combined protease extraction method for digestion on the surface of cultural relics has been developed. In 2009, Gabriella Leo et al. [47] used microwave-assisted trypsin to extract protein from the pigment layer of a 13th-century Italian cathedral dome fresco, and then used nano-LC-MS/MS and ESI to detect the extracted peptide segments. It was discovered that the binder was milk. Using this method to extract proteins will not harm cultural artefacts and is appropriate for samples with trace quantities, ageing, and degradation.

Recently, a novel hydrogel protein extraction material comprised of trypsin and chymotrypsin has been utilised in the analysis of cultural relic protein materials. Calvano et al. [48] directly coated the surface of an oil painting and analysed the peptides obtained after enzymatic hydrolysis using MALDI–TOF/TOF. They obtained a large number of casein, rabbit collagen, and egg yolk immunoglobulin peptide peaks, which increased protein coverage. It was discovered that bovine adhesive, eggs, and milk were used as binders when studying Italian statues painted in the 16th century using this technology. Using hydrogel to extract proteins prevents injury to the pigment layer; additionally, compared to samples treated with trypsin, more and shorter peptide fragments are obtained, which facilitates the identification of proteins based on the analysis of peptide fragments.

Ethyl-vinyl acetate (EVA) films are used to extract proteins from artefacts by coating the surface of cultural relic materials with moisture for 15–30 min. As follows is how EVA film is prepared: Vinyl acetate, strong anion, and strong cation exchange resin are fused and compressed to form a 150–200 m-thick film. This technology enables for the extraction of precious cultural relic materials in situ without damaging them. In addition, it leaves no residue on the surface of the artefacts. However, the preparation procedure is complex and requires specialised equipment, making it difficult to promote its current use. In 2017, Zilberstein et al. [49] utilised this method to conduct a proteomic analysis of the remnants on the margins of the manuscript of the renowned writer Mikhail Bulgakov. Proteins on the paper may be adsorbed by saliva or perspiration and left at the page's edge. Following the extraction of the EVA film, the protein is eluted, reduced, alkylated, and digested with trypsin. Proteins like periostein, N-acetyl--glucosaminidase, and norepinephrine were identified using liquid chromatography–tandem mass spectrometry (LC–MS/MS). All of these are biomarkers of kidney disease, proving that the author died from this condition.

3.2. Mass Spectrometry Analysis

Core to proteomics technology is the qualitative and quantitative analysis of extracted proteins by mass spectrometry. Chromatography and spectroscopy are the most comprehensive methods of protein analysis. Chromatographic techniques include gas chromatography (GC) [50], high-performance liquid chromatography (HPLC), and pyrolysis-gas chromatography (Py-GC), among others. Fluorescence spectroscopy [51], Fourier transform infrared spectroscopy (FTIR) [52], and laser Raman spectroscopy [53] are spectroscopic techniques. However, the disadvantages of chromatographic and spectroscopic techniques include their susceptibility to interference from other inorganic substances, their inability to characterise a single protein in mixtures of multiple proteins, and the similar amino acid compositions of proteins derived from different biological sources. Consequently, the chromatographic and spectroscopic techniques listed above are typically incapable of identifying species and are inconvenient for detecting protein mixtures.

With the continuous development of MS technology, proteomics research applications are expanding in both scope and depth. MS can detect amino acid sequences, post-translational modifications, and the molecular weight of peptides and proteins with precision. MS has become the most significant identification technology in proteomics [9,42,54,55] because of its high sensitivity, accuracy, and ease of automation. Triple quadrupoles, time-of-flight (TOF), and Orbitraps are the most prevalent types of mass analyzers. Triple quadrupoles have limited resolution, but their advantages are sensitivity and speed. TOF's ultra-high resolution and precision are more conducive to the identification of undiscovered species in complex situations. In addition, their selection is of superior quality. Due to the fact that MALDI generates molecules with a high mass-to-charge ratio (m/z), they are frequently employed in tandem with MALDI. Orbitraps have the highest resolution and mass accuracy and are coupled with ESI/nano-ESI due to their ability to resolve the numerous charge states. Each mass analyzer has distinct precision, sensitivity, and resolution, and can be chosen based on the detection needs [56]. MALDI-TOF-MS is able to obtain the relative molecular weights of numerous enzymatic peptide fragments and PMFS. It is suitable for large-scale rapid detection and is resistant to salt, detergents, and high-concentration buffer solutions. This technology has the following limitations: the peak expression of the obtained protein is frequently similar or overlapping; the number of different proteins cannot be determined; it has relatively low reproducibility and separation; it lacks stringing capability, thereby limiting further qualitative capabilities; and it requires precision instruments with high maintenance costs [57,58]. ESI-MS can be utilised to sequence peptides and determine the amino acid sequences of various peptides. Its benefits include broad solvent applicability, high ionisation efficiency, low sample volume requirement, and broad mass range [59]. Several disadvantages include: Capillaries can easily obstruct when detecting higher concentrations of samples or samples containing impurities; only highly volatile buffers (such as ammonium acetate) can be used; separate ESIs will interfere with random peaks when analysing the mixture; and the spectrum is too chaotic. QMS is appropriate for bottom-up methodologies, and its benefits include a simple structure, low cost, simple maintenance, high quantitative capability and sensitivity, and a wider linear range. There is no tandem polarity capability, insufficient qualitative capability, low resolution of fragment ions, and sluggish speed among the disadvantages.

MALDI-TOF-MS is capable of identifying not only individual proteins, but also composites and biomolecules. Due to the frequent presence of protein compounds derived from various biological sources (eggs, animal glue, milk, etc.) in cultural relics, MALDI-TOF-MS is employed for differentiation. For instance, mortar samples collected in 2009 from the Romanesque rotunda of Saint Catherine in Znojmo (Czech Republic) were analysed using MALDI–TOF MS. It can distinguish commonly used proteins (casein, collagen, egg protein) as additives and is used to determine mixtures and biological macromolecules. The results revealed the presence of two proteins in the mortar sample [60]: casein and collagen.

ESI-MS can generate ions with multiple charges, and molecules with a high molecular weight typically possess multiple charges. The distribution of charge states can precisely quantify molecular mass, thereby simultaneously providing accurate molecular mass and structural information. Therefore, it can be used for the analysis of organic macromolecules. Yoko Taniguchi [61] examined the Bamiyan Wall Paintings' binders in 2022. Using ESI-MS/MS for protein analysis, researchers detected type I collagen in cattle, which is distinct from collagen from swine and horses, as well as type III collagen from cattle. This not only validates the previous species detection results, but also suggests that the material used is bovine skin as opposed to bovine bone because the abundance of this protein in bones and tendons is significantly higher than in skin. Consequently, ESI-MS detection allows for the retrieval of more information regarding the protein materials used in cultural relics, leading to more in-depth findings.

The QMS technique has higher ion and protein fractions and is both rapid and sensitive. Stepanka Kuckova [62] analysed four distinct protein additives (animal glue, blood, egg, and milk) added to aged mortar samples using LC-ESI-Q-TOF MS. LC-ESI-Q-TOF MS was able to identify milk additives that could not be identified by MALDI-TOF MS, as it was able to identify more specific peptides in the peptide mixture.

Several mass spectrometry techniques can be used in series to detect peptides with various physical and chemical properties in order to improve the ability to detect proteins in cultural relic materials. Proteins can be extracted from low-volume samples without the need for protein hydrolysis. Using a combination of MALDI-TOF-MS, LC-MS/MS, and

LTQ-Orbitrap, Tripkovic T. et al. [63] identified 19th-century Eastern Orthodox paintings in 2015. Various methods were used to identify distinct peptides of the same protein in this investigation. Using LC-MS/MS with an ESI-LTQ-Orbitrap, less abundant proteins were identified compared to MALDI-TOF MS/MS. Since only a few peptides elute at once, chromatographic separation in front of the ESI source enabled the identification of multiple proteins. MALDI-TOF MS/MS has higher throughput and shorter analysis periods, whereas HPLC-ESI-LTQ-Orbitrap provides more detailed results and detects proteins at lower concentrations. This combination of ionisation techniques allowed for overlap and compensation.

3.3. Biological Information Database Retrieval

The third essential component of proteomics technology is bioinformatics research, the most important phase of which is the creation of a database. Researchers can identify the source species of the protein by comparing the MS data obtained after the aforementioned analyses with the database data. Completeness and breadth of the database are essential for confirming the labelled peptides and identifying species. If ancient samples lack sequence information in the database or if the sequence information is incomplete, modern samples can be used as a reference. In 2007, Kuckova et al. [62] compiled a database of modern samples of commonly used binders (egg yolks, egg whites, casein, milk, curds, whey, gelatine, and various types of animal glue); then, the programme Mass-2.0-alpha4 simulated the decomposition of the aforementioned substances and compared this to the amino acid sequence of a single protein in a database (such as ExPASy). The outcomes matched. The peptide mixture obtained from a few micrograms of pigment layers from Edvard Munch paintings was then analysed using MALDI–TOF MS, which revealed that eggs were the binder. Dallongeville et al. [64] utilised proteomics techniques based on Fourier transform ion cyclotron resonance MS (FTICR-MS), nano-LC, nano-ESI, high-resolution MS (HR/MS), and MS/MS to identify sequence information that was not in the protein database in 2011. Animal gels from several species (bovine, rabbits, and fish) that are commercially available were analysed, and the specific peptides of fifteen species of cattle, three species of rabbit, and three species of fish were identified.

Subsequent analyses revealed that the complex simulation materials consisted of lead white, animal adhesive, and linseed oil; the source species of the animal glue species was also identified. Last but not least, the technology was successfully applied to a 50 g gold-plated sample of a church from the 18th century, demonstrating that the adhesive glue used for plaster brackets and gold foil was bovine glue. Azémard et al. [65] utilised electrospray quadrupole time-of-flight (ESI-Q-TOF) MS in conjunction with ultra-performance liquid chromatography-MS (UHPLC-MS) and LC–MS/MS to identify the species of ancient fibres in Xinjiang in 2019. Due to the arid climate of the Tarim Basin, where the sample is located, the area is ideal for preserving animal fibres. The difference in keratin deamidation rates between ancient and contemporary samples is negligible. By analysing the keratin-specific peptides of ancient samples from this region for species differentiation, the proteomic database of extinct species can be expanded. The results showed that goats, sheep, and cattle were the primary sources of fibre for textiles. In 2021, Dubrovskii et al. [66] proposed a broadband collision-induced dissociation (bbCID) model, which was used to acquire characteristic peptides of ovalbumin and collagen for database construction. After testing the samples with HPLC-MS and nano-LC-MS, the protein was identified by comparing it to the database. This method was used to analyse building samples from the 19th century. Collagen was identified as the primary component using six characteristic peptides. This technology can test samples with multiple layers, and there is no need to re-prepare samples for subsequent analyses.

3.4. Characteristics of Proteomics Technology

Proteomics technology is a high-sensitivity, high-resolution, and high-throughput detection technology that enables the analysis and detection of a dozen complex samples simultaneously; it can detect with only a few micrograms of samples, and numerous sampling methods that do not require the destruction of cultural relic materials have been developed. In addition, the detection limit is low, the tested samples can be repeatedly applied to other detection methods, making it appropriate for samples of valuable cultural relic materials, and the distinction between protein-derived animals can be accurate to the species level. However, the primary disadvantage of this technology is that samples typically require complex pre-treatment procedures, such as separation, purification, and hydrolysis in inorganic solutions, which reduces the amount of information pertaining to biological sources. In addition, the intricate pre-treatment procedure increases the likelihood of sample loss and contamination. Complex instruments impose stringent requirements on experimental platforms and operators. They are only appropriate for professional scientific research institutions and cannot be analysed and examined by various cultural relic material preservation units.

4. ELISA Technology

4.1. The Concept of Enzyme-Linked Immunosorbent Assay

ELISA is a labelled immunoassay technique that combines the specific binding reaction of antigen–antibody with the color-rendering reaction of enzymes and catalytic substrate to enhance sensitivity [3] by amplifying the signal. In a published article promoting the establishment and development of ELISA [67], Swedish researchers Engvall and Perlmann first proposed the use of ELISA for the quantitative detection of antibodies during the 1970s. The substrate is added after the enzyme is labelled on the antibody, the antigen is immunologically bound to the enzyme-labeled antibody, and then the antigen is specifically bound to the enzyme-labeled antibody. The product of the colour reaction between the substrate and the enzyme is coloured. The colour intensity of the product correlates positively with the quantity of antigen or antibody in the test substance. Using an enzyme marker, the absorbance of the product at a particular wavelength can be measured, and the antigen or antibody can be quantitatively analysed. An antigen is a molecule that can stimulate an organism's immune system, and an antibody is a glycoprotein that can bind to an antigen in a specific manner. Proteins can be used as an antigen to generate antibodies, and for immune experiments to generate specific antibodies, only the protein must be extracted and purified.

The direct method, double antibody sandwich method, indirect method, and competition method are typical ELISA detection techniques used for protein analysis of cultural relic materials [68] (Figure 2). The direct method involves adsorbing the antigen onto a solid-phase carrier, incubating it at an appropriate temperature and relative humidity, and then washing away any unbound antigens and impurities. To bind other unbound sites on the solid-phase carrier, a high concentration of nonspecific proteins and enzyme-labeled specific antibodies are added. At an optimal temperature and relative humidity, the antigen and antibody react completely, and the sample is then rinsed to remove unreacted antibodies. Finally, the chromogenic substrate is added, and within a certain amount of time, the enzyme-catalyzed colour develops. The experimental operation requirements for the labelling reaction of antibodies are high; each antigen detected by the direct method must be labelled with a specific antibody that interacts with it, resulting in a high cost of detection. Using the sandwich method, a known antibody is affixed to the surface of a solid-phase carrier, followed by the test sample. If the sample contains an antigen, specific binding will occur; then, an enzyme-labeled antibody is added to produce an antibody–antigen–enzyme-labeled antibody "sandwich" structure. After the addition of the substrate, the antigen in the sample is detected and analysed using a color-generating reaction. Double antibody sandwiching is a common technique for detecting macromolecular antigens [69]. However, when using the double antibody sandwich method to detect each antigen, the specific antibody that reflects that antigen must be labelled; when combined with indirect methods, each antigen requires two antibodies, making the double antibody sandwich method more expensive. The indirect technique can be used to detect antigens. The principle is to adsorb the antigen to be tested onto the solid phase, followed by the addition of specific antibodies and enzyme-labeled antibodies. Alternately, after the antigen has been adsorbed in the solid phase, the antibody to be examined and the enzyme-labeled antibody are successively added. In the indirect method, the colour intensity is directly proportional to the concentrations of the coated antigen and the primary antibody. The binding between the first and second antibodies is based on the principle that cross-reactions can occur between distinct subclasses of the same antibody. Therefore, specific antibodies can be applied directly to ELISA without being labelled, preserving the high activity of the specific antibody, reducing the number of experimental steps, and saving time and money. Consequently, indirect methods are the most popular. It is also the most popular method for detecting proteins in cultural relics [70]. The competition method is distinguished by the interaction between the antigen being tested and the enzyme-labeled antigen and the solid-phase antibody. The more the enzyme-labelled antigen binds to the solid-phase antibody, and the darker the colour, the lower the antigen concentration in the test sample. The intensity of the colour of the competitive solution is directly proportional to the antigen concentration in the mixed solution and inversely proportional to the antigen concentration on the solid carrier. This method is primarily used to determine small-molecule hapten [71–73].

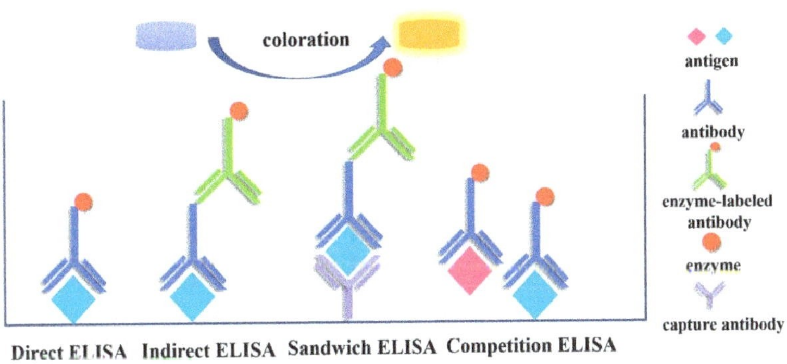

Figure 2. Common detection methods of ELISA.

4.2. Application of Early ELISA in Protein Detection of Cultural Relics Materials

Since the introduction of ELISA technology, it has been widely used for the detection of viruses, antibiotics, heavy metal ions, disease-related proteins, residues, etc., due to its high sensitivity, easy and rapid operation, the small amount of sample required, low cross-reactivity, and high specificity. The development of ELISA technology has also piqued the interest of archaeologists, who have started applying it to the detection of proteins in materials from cultural relics. Earlier studies on the application of ELISA in archaeology were numerous. In 1990, for instance, ELISA was used to detect albumin protein in ancient corpses and animal remains [74]. It was also used to determine that the pigment deposits of the Chumash Indians in the United States consisted of both animal and human blood. Cattabeo [75] utilised ELISA to analyse ancient human bone extracts in 1992. Albumin was discovered in 23 of the 31 skeletons, whereas only one skeleton contained IgG. Therefore, it has been demonstrated that albumin is a significantly superior target molecule for long-term survival in ancient bones. The results demonstrated that there is no cross-reaction between human and animal materials when using ELISA to detect albumin in bones, and that as little as 10 ng of protein is detectable, thereby expanding the possibilities for bone archaeology. In 2002, Schweitzer [76] used ELISA to identify immunologically active polypeptides in 100,000 to 300,000-year-old fossilised skulls. This demonstrates that ELISA is sensitive enough to detect proteins that are extremely aged.

4.3. ELISA Detection of Proteins in Cultural Relics Materials in a Complex Environment

With the ongoing development and enhancement of ELISA, a large number of research results for protein analysis and detection in materials from cultural relics have emerged, and the sensitivity and specificity of detection have improved continuously. In the protein processing method, ELISA employs the trypsin and other prior procedures of aged proteins so that the protein's antigen determinant is exposed and can react with antibodies. This method detects ageing and denatured proteins in cultural relic materials without the need for a complex protein separation and purification procedure [45]. Protein components that must be detected in cultural relic materials frequently exist in complex environments and are difficult to separate; for instance, protein-cemented materials in mural layers are frequently mixed with inorganic materials, and cations of lead, copper, calcium, and iron are present in pigment layers. When amino acid analysis is performed using mass spectrometry, it will interfere with the derivatization process and alter the results [77]. In the presence of impurities, such as inorganic materials, active cations, and organic polymers, ELISA is superior to other detection methods for identifying proteins in cultural relic materials. In 2010, Palmieri et al. [78] analysed proteins in samples of cultural relics materials containing inorganic matrices using ELISA. After testing samples of 13th-century murals, it was determined that casein could still be detected in common mural substrates, such as carbonated stucco or gypsum, with a detection limit as low as 1 nanogram, indicating that these inorganic matrices will not affect the detection ability and sensitivity of ELISA. Similarly, the use of ELISA technology for determining the age of protein binders in pigments containing distinct metal ions will not affect the test results. In 2015, Lee et al. [77] modified the colour portrayal of Alkaline Phosphatase (AP), a commonly used enzyme in ELISA, to horseradish peroxidase (HRP), which is more sensitive than AP and has a greater magnification range. Animal cements containing pigments such as lead white and ultramarine are mined for proteins. The ELISA signal remains unaffected, and the protein type can still be identified. Similarly, a sample was extracted from a 12th- to 13th-century Peruvian feather robe. The ELISA results revealed that the sample's adhesive was arabino-galactose gum, not animal glue as was previously believed. Type I collagen was detected in paper samples obtained from 19th-century to early 20th-century watercolour paintings. Ultimately, type I collagen and ovalbumin were discovered in tempera paintings on canvas from the 14th century. These results demonstrate that ELISA is capable of detecting various classes of proteins containing distinct metal ions.

4.4. ELISA Detection of Trace Samples

ELISA can detect and identify proteins in cultural relic materials that are stored in adverse environments, severely damaged, ageing, or even completely denatured, as well as those that are difficult to distinguish visually. It can also differentiate between protein types, determine their biological origin [79] and detect and identify even trace concentrations of proteins. Liu [80] used indirect ELISA and indirect competitive ELISA to detect and identify samples of wool and leather from cultural relics. Cowhide was identified as the variety of leather found in samples from dry regions of ancient Xinjiang. The ELISA indirect detection threshold for keratin was 10 ng/mL. Additionally, the collagen type I in the three ancient leather samples was identified, and the results of various proportions were obtained, thereby completing the species identification. Wu [81] laboured in 2017 to detect lacquer in samples of cultural relic material. Due to the complex composition of the remaining samples, which were contaminated and contained only trace quantities, the FTIR and Py-GC/MS results for lacquer phenol were insufficient. The glycoproteins in the dried lacquer films were analysed using ELISA, and the results were applied to eight samples collected from six remnants or ancient structures in different Chinese cities. This study's detection limit reached 106 g/mL, and muddy samples were also analysed. In 2021, Weng [82] used ELISA to analyse samples of Chinese building mortars from 4300 years ago (approximately 2300 BC) to determine if they contained traces of commonly used binders, such as glutinous rice, tung oil, sugar, and animal adhesive. The results indicated that the samples contained components of animal adhesive. ELISA was able to detect protein components in old, exceedingly low-content mortars.

5. The Combination of ELISA and Other Technologies

There have been an increasing number of techniques in recent years to enhance ELISA's detection effectiveness by fusing it with other technologies. In 2016, Liu [83] employed immunofluorescence microscopy (IFM) and ELISA to pinpoint the precise location of protein binders in the mural layer of the Xumi mountain caves and determine their composition. Animal glue and egg whites, two frequent sticky ingredients found in ancient Chinese paints, were found. IFM can characterise the structure distribution information of each layer of binder materials because it is a multi-layer structure that is challenging to mechanically peel off, and ELISA combined with IFM has high detection sensitivity. The protein components in it can still be detected and classified when the cemented material is mixed with the pigment.

Immune-labeled magnetic beads can bind protein more effectively and withstand interference from impurities when used in conjunction with ELISA. The magnetic beads are appropriate for use at archaeological sites and can be mass-produced as well. Zhang Wei [84] created an ELISA assay for the detection of silk protein in 2020 using immune-labeled magnetic beads. Silk protein was used to adsorb immune-labeled magnetic beads, and following adsorption, the silk protein was eluted for indirect ELISA testing. Testing of soil samples from the Qingtai site in Zhengzhou, which had evidence of probable silk breakdown, revealed the presence of silk protein there. Because immune-labeled magnetic beads have good durability, this approach offers higher sensitivity than standard ELISA. Its portability and strong anti-interference ability make it suited for trace protein identification in ancient sites. Table 2 provides a summary of the detection methods and outcomes for samples of historical proteins.

Table 2. Detection techniques and results of ancient protein samples.

	Sample	Country of Sample	Time	Detection Method	Detected Protein
Proteomics	Cementitious Materials in Renaissance Art Painting [46]	Italy	2006	MALDI-TOF, nano-LC, nano-ESI, Q-q-TOF-MS, MS	Ovalbumin
	17th Century Painings [62]	Norway	2007	MALDI-TOF MS	Egg protein
	Cathedral dome painting binder [47]	Italy	2009	nano-LC-MS/MS; ESI	Casein
	Mortar sample for Romanesque rotunda [60]	Czech Republic	2009	MALDI-TOF-MS	Casein; Collagen; Egg protein
	18th century church binder [64]	France	2011	FTICR-MS, nano-LC, nano-ESI, HR MS, MS/MS	Bovine collagen
	Orthodox paintings [63]	Serbia	2015	LTQ-Orbitra, MALDI-TOF MS/MS	Yolk protein
	The manuscript of the famous writer Mikhail Bulgakov [49]	Russia	2017	LC-MS/MS	Periostein; acetyl-β-glucosaminidase; Norepinephrine
	Ancient fibers [65]	China	2019	ESI-Q-TOF, UHPLC-MS, LC-MS/MS	Goat keratin; Sheep keratin; Sovine keratin
	Handmade Papers (13th–15th century) [45]	Italy	2020	SDS-PAGE	Collagen
	Handmade Papers (13th–15th century) [45]	Italy	2020	SDS-PAGE	Collagen
	Painted statues [48]	Italy	2020	MALDI-TOF/TOF	Casein; Rabbit collagen; Egg yolk immunoglobulin
	19th-century buildings [66]	Russia	2021	HPLC–MS, nano-LC–MS	Collagen
	Wall Paintings [61]	Afghanistan	2022	ESI-MS/MS	Type I collagen Type III collagen

Table 2. *Cont.*

	Sample	Country of Sample	Time	Detection Method	Detected Protein
ELISA	13th Century Murals [78]	Italy	2010	ELISA	Casein
	Peruvian feather robe from the 12th–13th centuries [77]	Peru	2015	ELISA	Arabino-galactose gum
	Watercolor paintings from the 19th century to the early 20th century [76]	America	2015	ELISA	Type I collagen
	14th-century tempera paintings [77]	Italy	2015	ELISA	Type I collagen; Ovalbumin
	Wool and leather [80]	China	2016	ELISA	Type I collagen; Keratin
	Murals binders [83]	China	2016	ELISA, IFM	Egg protein; Collagen
	Dried lacquer films [81]	China	2017	ELISA	Glycoproteins
	Soil samples [84]	China	2020	IMB, ELISA	Silk protein
	Building mortars 4300 years ago [82]	China	2021	ELISA	Collagen
Combination of the two methods	Painted wood panels from the 13th century [85]	China	2018	ToF-SIMS, Dot-ELISA	Casein; Rabbit collagen; Egg protein
	Silks [13]	China	2019	LC–MS/MS, ELISA	Sericin; Silk fibroin
	Silk residues [86]	China	2021	IMB, ELISA, LC–MS/MS	Silk protein

5.1. Characteristics of Enzyme-Linked Immunosorbent Assay

The minimal sample purity requirements, straightforward sample pre-treatment, limited number of experimental procedures, lack of need for complicated instruments, ease of operation, and low cost of ELISA all contribute to its high sensitivity and low sample volume requirement. Therefore, it is appropriate for the initial classification of different cultural relic elements at archaeological sites that contain proteins. It is also appropriate for the quick and simple identification of cultural heritage artefacts with mixed proteins since protein detection is selective and non-target proteins have less of an impact on target proteins. It can be widely employed by cultural and insurance sites with various situations because it has few criteria for experiment instruments and laboratory workers. Because it is impossible to distinguish between distinct species of animals belonging to the same genus, ELISA can only accurately identify the species source of proteins at the genus level. Additionally, because ELISA cannot detect many substances simultaneously, molecules with related structures may react with one another.

5.2. Combined Use of Proteomics and ELISA

Proteomics and ELISA have increasingly been used in research in recent years to analyse proteins. Proteomics approaches have high throughput and accuracy benefits, and the evaluated samples can be reused for ELISA testing. Low sample purity requirements, easy operation without complicated equipment, low cost, and little influence of non-target proteins on the results are all characteristics of ELISA. The composition and source of the sample can be correctly recognised by combining the benefits of the two analysis techniques (Table 3). In order to characterise the organic adhesives employed in painted wood panels from the 13th century, two techniques were applied in 2018: time of flight secondary ion mass spectrometry (ToF-SIMS) and Dot-ELISA [85]. While ELISA has great specificity and sensitivity and can perform a single identification of proteins in trace samples, ToF-SIMS offers high spatial resolution and high surface sensitivity. The outcomes of both techniques demonstrated the existence of a rabbit glue and milk mixture in the paint layer of the sample while ruling out the usage of eggs. Proteins can be qualitatively analysed and their species identified using immunoassay and proteomics techniques.

Table 3. Comparison of characteristics of two detection methods.

	Proteomics	ELISA
Sample weight	μg, non-destructive sampling	mg
Pre-treatment process	complex	simple
Number of samples for single test	more	less
Instrument	precision	simple
Operating procedure	complex	simple
Inspection cost	high cost	low cost
Sample recovery	recyclable	non-recyclable
Limit of detection	10^{15} (mol/L)	10^{9} (mol/L)
Species identification precision	species	genus
Whether or not false negative	yes	no

Sericin and silk fibroin were successfully identified as marker proteins in 2019 by Chen [13] using LC-MS/MS to identify silkworm cocoon protein. The heavy chain of the mulberry silk protein's distinctive peptides were tested as haptens, while the tussah cocoon's silk protein was used as a full antigen. Through animal immunology, specific antibodies to various silks were developed, and distinct species' silk proteins were successfully identified. In 2021, Zheng et al. [86] created a technique for quick enrichment and detection of silk residues based on immune-labeled magnetic beads and ELISA. The detection threshold was at 5.12 ng/mL. After that, they employed LC-MS/MS to further validate the findings. The study demonstrates that the composition and source of silk residues may be precisely determined by combining immunological and proteomic approaches.

5.3. Expectation

Proteomics technology and ELISA have gradually been used to analyse the proteins in cultural relic materials in recent years. This has not only satiated the demands for low destructiveness, high resolution, high throughput, and high sensitivity, but also revealed information about the biological sources of proteins. To increase the effectiveness of protein extraction, proteomics technology must further optimise the extraction process. Additionally, it is impossible to compare many significant spectra with the database. To identify the iconic proteins and create a more comprehensive database of the proteins found in cultural relic materials, it is important to examine the proteins in additional samples of cultural relic materials. The accuracy and specificity of biological source identification will keep rising thanks to the ongoing development of bioinformatics, databases, and analysis software. The future of this field lies on combining conventional ELISA technology with new technologies for extraction and detection. For instance, ELISA with immune-labeled magnetic beads can be used to detect minute quantities of proteins. ELISA and other sensors can also be used to create biosensors, such as electrochemical immunosensors. These techniques enable successful qualitative and quantitative studies, they are simple to use, they can be automatically recognised, they can achieve higher sensitivity, and they can identify proteins in a larger range of materials used to create cultural relics. Additionally, because protein is reversible, has strong biocompatibility, and the non-changing morphology, content, and qualities of cultural relic materials, it has steadily been employed as a bioremediation material in the restoration and reinforcement of such materials. In the future, this could develop into an improved repair and protection strategy for cultural heritage items. Finally, the methods for analysing protein in materials from cultural relics will considerably improve with the advancement of protein analysis technology.

Author Contributions: Writing—original draft preparation, Q.W. writing—review and editing, B.Z. and Y.H. All authors have read and agreed to the published version of the manuscript.

Funding: This research was funded by [National Key R&D Program of China] grant number [No. 2022YFF0903802].

Institutional Review Board Statement: Not applicable.

Informed Consent Statement: Not applicable.

Data Availability Statement: No new data were created or analyzed in this study. Data sharing is not applicable to this article.

Conflicts of Interest: The authors declare no conflict of interest.

References

1. Maria, G.G.; Roberto, M.; Enrica, P. Back to the past: Deciphering cultural heritage secrets by protein identification. *Appl. Microbiol. Biotechnol.* **2018**, *102*, 5445–5455.
2. Xu, L. Study on the Comprehensive Analysis Method of Traditional Mortar Materials. Master's Thesis, Zhejiang University, Hangzhou, China, 2018.
3. Hu, W.J. Immunoassays Technology of Binding Media in Precious Ancient Polychrome Cultural Relics. Master's Thesis, Zhejiang University, Hangzhou, China, 2016.
4. Chen, H.F.; Gong, D.C.; Huang, W.C.; Liu, B. SDS-PAGE Analysis of Aging Characteristics of Liao Dynasty Silk Unearthed in Yemaotai, Faku, Liaoning. *Sci. Conserv. Archaeol.* **2010**, *22*, 9–13.
5. Wang, K.; Hu, D.B. Hydroxyapatite: Collagen Biomimetic Composite Material in Conservation of Tortoise Shell Relics. *JNMCH* **2013**, *3*, 141–152.
6. Zhang, Y.; Chen, Z.; Liu, X.; Shi, J.; Chen, H.; Gong, Y. SEM, FTIR and DSC Investigation of Collagen Hydrolysate Treated Degraded Leather. *J. Cult. Herit.* **2021**, *48*, 205–210. [CrossRef]
7. Pasian, C.; Secco, M.; Piqué, F.; Artioli, G.; Rickerby, S.; Cather, S. Lime-based injection grouts with reduced water content: An assessment of the effects of the water-reducing agents ovalbumin and ethanol on the mineralogical evolution and properties of grouts. *J. Cult. Herit.* **2018**, *30*, 70–80.
8. Wang, Z.Y. Proteomics Analysis and Ultra-Sensitive Detection of Archaeological Hairs. Master's Thesis, Zhejiang Sci-Tec University, Hangzhou, China, 2021.
9. Wang, J.; Li, J.; Chao, X.; Chen, Y.; Huang, Y.; Mai, B.; Li, Y.; Cao, J. The influence of the photothermal aging exerting on the GC-MS identification of the common proteinaceous binding media in polychromy artworks. *JNWU* **2020**, *50*, 595–605.
10. Liu, L.Y. Multi-Analytical Approach to the Study of Pigments and Binding Media in Precious Ancient Polychrome Cultural Relics. Master's Thesis, Zhejiang University, Hangzhou, China, 2017.
11. Abelson, P.H. *Paleobiochemistry: Organic Constituents of Fossils*; Yearbook, No. 531954; Carnegie Institution of Washington: Washington, DC, USA; pp. 97–101.
12. Li, J.J.; Zhang, B.J. Research status of binders in ancient painted cultural relics—Based on data analysis of research papers from the Web of Science. *Sci. Conserv. Archaeol.* **2019**, *31*, 119–129.
13. Chen, R.R. Characterization of Ancient Silk Fabrics Based on Proteomics and Immunological Techniques. Master's Thesis, Zhejiang Sci-Tec University, Hangzhou, China, 2020.
14. He, Q.J. Study on the Role of Sizing Agent in Ancient Calligraphy & Painting and the Development of Neutral Aluminum Salt Sizing Precipitator. Ph.D. Thesis, Northwest University, Xi'an, China, 2019.
15. Chi, H.H. Artificial Aging and Consolidation of Silk Painting Colored by Mineral Pigment. Ph.D. Thesis, Zhejiang Sci-Tec University, Hangzhou, China, 2020.
16. He, Q.J.; Wang, L.Q. Study on application of modern instruments analysis in identification of painting relics binding medium. *China Adhes.* **2007**, *19*, 19–22.
17. Deng, Z.M.; Jiang, Z.H.; Qin, D.C. Evaluation on the moisture-proof performance of bamboo coated by lacquer in hot and humid environment. *China For. Prod. Ind.* **2015**, *42*, 14–16.
18. Wang, C.; Qi, L.X.; Yang, J.; Zhou, W.; Dang, X. Status of research on the conservation of bone and antler relics. *Sci. Conserv. Archaeol.* **2016**, *28*, 118–125.
19. Wang, J. Research on the Phenomenon of Decline in the Solubility of Silk Protein in Silk Cultural Relics and Its Solution. Master's Thesis, Shandong University, Jinan, China, 2021.
20. Zhang, B.J.; Hu, W.J.; Zhang, K. Research on the Egg White in Ancient Tabia Mortar by Enzyme-Linked Immunosorbent Assay (ELISA). *J. Build. Mater.* **2015**, *18*, 716–720.
21. Wilson, J.; van Doorn, N.L.; Collins, M.J. Assessing the extent of bone degradation using glutamine deamidation in collagen. *Anal. Chem.* **2012**, *84*, 9041–9048. [CrossRef]
22. Solazzo, C.; Wilson, J.; Dyer, J.M.; Clerens, S.; Plowman, J.E.; Von Holstein, I.; Collins, M.J.; Walton Rogers, P.; Peacock, E.E. Modeling deamidation in sheep α-keratin peptides and application to archeological wool textiles. *Anal. Chem.* **2014**, *86*, 567–575. [CrossRef] [PubMed]
23. Xie, J. Thermal aging Mechanism and Soften Technology of Leather. Master's Thesis, Zhejiang Sci-Tec University, Hangzhou, China, 2016.
24. Zhang, M. Study on Collagen Extraction Methods from Degraded Leather Cultural Relics. Master's Thesis, Shandong University, Jinan, China, 2021.

25. Pankin, D.; Povolotckaia, A.; Borisov, E.; Rongonen, S.; Mikhailova, A.; Tkachenko, T.; Dovedova, N.; Rylkova, L.; Kurochkin, A. Investigation of wafers used as paper binding in the academician von Struve manuscripts. *J. Cult. Herit.* **2021**, *51*, 125–131. [CrossRef]
26. Zhao, L. Tracing the animal origin of organic cultural relics in the Palace Museum through mitochondrial DNA technology. *China Cult. Herit. Sci. Res.* **2019**, *3*, 75–82. (In Chinese)
27. Paolo, C.; Antonella, C. Enzymes as tools for conservation of works of art. *J. Cult. Herit.* **2021**, *50*, 73–87.
28. Banks, P.N. Paper Cleaning. *Restaur. Int. J. Preserv. Libr. Arch. Mater.* **1969**, *1*, 52–66. [CrossRef]
29. Frantisek, M. Enzymatic consolidation of a painting: Seventeenth century landscape from skokloster palace. *Stud. Conserv.* **1982**, *27*, 135–138.
30. Paolo, A.; Giacomo, Z.; Pamela, A. Art-loving bugs: The resurrection of Spinello Aretino from Pisa's cemetery. *Proteomics* **2005**, *5*, 2453–2459.
31. Zhou, Y. The Discovery of Silk in the Ritual Pit of Sanxingdui and its Significance. *Chin. LH* **2021**, *12*, 37–48.
32. Su, B.M.; Zhang, A.M.; Hu, Z.M.; Li, Z.X. Application of Chromatographic Method in the Analysis of Cementitious Materials in Ancient Paintings. *Dunhuang Res.* **2000**, *1*, 82–86.
33. Chen, D.M.; Si, C.D.; Long, S.J. Research Progress of Spectral Technologies of Binging Media Used in Paintings. *Spectrosc. Spect. Anal.* **2020**, *40*, 961–966.
34. Zhao, Q.; Zhang, L.H.; Zhang, Y.K. Recent Advances in Proteomics. *Chin. J. Appl. Chem.* **2018**, *35*, 977–983.
35. Candiano, G.; Bruschi, M.; Musante, L.; Santucci, L.; Ghiggeri, G.M.; Carnemolla, B.; Orecchia, P.; Zardi, L.; Righetti, G. Blue silver: A very sensitive colloidal Coomassie G-250 staining for proteome analysis. *Electrophoresis* **2004**, *25*, 1327–1333. [CrossRef]
36. Jennifer, A.; Milana, P.; Ei, W.W. Comparison of protein expression levels and proteomically-inferred genotypes using human hair from different body sites. *Forensic. Sci. Int.-Gen.* **2019**, *41*, 19–23.
37. Xin, P.; Kuang, H.-X.; Li, X.-L.; Wang, Y.; Zhang, B.-M.; Bu, H.; Wang, Z.-B.; Meng, Y.-H.; Wang, Y.-H.; Wang, Q.-H. Proteomics and its application to determine mechanism of action of traditional chinese medicine. *China J. Chin. Mater. Med.* **2018**, *43*, 904–912.
38. Yang, Q.; Wang, D.; Chang, L.L.; Sun, Y.; Jin, X.; Yang, X. Progress in Mass Spectrometry and its Application in Proteomics. *Chin. Agric. Sci. Bull.* **2015**, *31*, 239–246.
39. Gong, P.T.; Xu, R.S.; Fang, X.J. Proteogenomics: Progress, Strategy and Problem. *Genom. Appl. Biol.* **2014**, *33*, 1169–1180.
40. Li, N.; Wu, S.-F.; Zhu, Y.-P.; Yang, X.-M. The Progress of Protein Quality Control Methods in Shotgun Proteomics. *Prog. Biochem. Biophys.* **2009**, *36*, 668–675. [CrossRef]
41. Zhao, J.; Wang, H.W.; Tian, E.J. Proteomics Experiment Technologies and Their Applications. *Prog. Vet. Med.* **2015**, *36*, 116–120.
42. Yu, Z.N.; Zhang, Y.; Huang, X.W. Application of proteomics in diagnosis and pathogenesis of genetic metabolic diseases. *Chin. J. Lab. Med.* **2022**, *45*, 300–304.
43. Zhang, W. Progress in Mass Spectrometry Acquisition Approach for Quantitative Proteomics. *Chin. J. Anal. Chem.* **2014**, *42*, 1859–1868. [CrossRef]
44. Radovan, H.; Stepanka, K.; Janka, H. Matrix-assisted laser desorption/ionization time-of-flight mass spectrometry as a tool for fast identification of protein binders in color layers of paintings. *RCM* **2004**, *18*, 1896–1900.
45. Proietti, N.; Roselli, G.; Capitani, D.; Pettinari, C.; Pucciarelli, S.; Basileo, S.; Scognamiglio, F. Characterization of handmade papers (13th–15th century) from Camerino and Fabriano (Marche, Italy). *J. Cult. Herit.* **2020**, *42*, 8–18. [CrossRef]
46. Tokarski, C.; Martin, E.; Rolando, C.; Cren-Olivé, C. Identification of proteins in renaissance paintings by proteomics. *Anal. Chem.* **2006**, *78*, 1494–1502. [CrossRef]
47. Leo, G.; Cartechini, L.; Pucci, P.; Sgamellotti, A.; Marino, G.; Birolo, L. Proteomic strategies for the identification of proteinaceous binders in paintings. *Anal. Bioanal. Chem.* **2009**, *395*, 2269–2280. [CrossRef] [PubMed]
48. Calvano, C.D.; Rigante, E.C.; Cataldi, T.R.; Sabbatini, L. In situ hydrogel extraction with dual-enzyme digestion of proteinaceous binders: The key for reliable mass spectrometry investigations of artworks. *Anal. Chem.* **2020**, *92*, 10257–10261. [CrossRef] [PubMed]
49. Zilberstein, G.; Maor, U.; Baskin, E.; D'Amato, A.; Righetti, P.G. Unearthing Bulgakov's trace proteome from the Master i Margarita manuscript. *J. Proteom.* **2017**, *152*, 102–108. [CrossRef]
50. Castellani, L.; Ferrantelli, P.; Sinibaldi, M.; Vigliano, G. Identification of proteinaceous adhesives in the wooden backing of Piero della Francesca's painting Pala of Saint Bernardino: A gas chromatographic study. *J. Cult. Herit.* **2001**, *2*, 209–215. [CrossRef]
51. Chirco, G.; Portale, E.C.; Caponetti, E.; Renda, V.; Martino, D.F.C. Investigation on four centuripe vases (late 3rd-2nd cent. B.C.) by portable X-ray fluorescence and total reflectance-FTIR. *J. Cult. Herit.* **2021**, *48*, 326–335. [CrossRef]
52. Gomaa, A.M.; Mostafa, A.H.; Hany, A. Analytical techniques used for condition assessment of a late period mummy. *J. Cult. Herit.* **2005**, *48*, 83–92.
53. Edwards; Howell, G.M.; Chalmers, J.M. Raman spectroscopy in archaeology and art history. *J. Raman Spectrosc.* **2004**, *35*, 607.
54. Ying, C.F.; Qing, X.; Shu, Z. Characterization and quantitation study of ancient lacquer objects by NIR spectroscopy and THM-Py-GC/MS. *J. Cult. Herit.* **2020**, *46*, 95–101.
55. Lu, R.; Kamiya, Y.; Miyakoshi, T. Applied analysis of lacquer films based on pyrolysis-gas chromatography/mass spectrometry. *Talanta* **2006**, *70*, 370–376. [CrossRef]
56. Ji, M.C.; Fu, B.; Zhang, Y.J. Recent Progress of Analytical Methods of Proteomics Based on Mass Spectrometry. *J Chin. Mass Spectrom. Soc.* **2021**, *42*, 862–877.

57. Yan, X.Y.; Gao, J.Z. Application of MALDI-TOF MS in early diagnosis of non-small cell lung cancer. *J. Med. Theory Pract.* **2022**, *35*, 1478–1481.
58. Wang, Y. Quantitative Analysis of Three Kings of Mycotoxins in Distillers Dried Grains with Soluble by Ultral-High Performance Liquid Chromatography-Tandem Mass Spectrometry. Master's Thesis, Jilin University, Jilin, China, 2018.
59. Li, J.; Ren, Q.L.; Pan, Y.J. Mass Spectrometric Studies on the Addition Reaction of HCCA and Sulfhydryl-Containing Living Substance in MALDI-TOF MS Analysis. *J Chin. Mass. Spectrom. Soc.* **2022**, *43*, 133–141.
60. Stepanka, K.; Radovan, H.; Milan, K. Application of peptide mass mapping on proteins in historical mortars. *J. Cult. Herit.* **2009**, *10*, 244–247.
61. Taniguchi, Y.; Kawahara, K.; Takashima, M.; Cotte, M.; Mazurek, J.; Kumazawa, Y.; Taga, Y.; Nakazawa, T. Organic Materials Used for Giant Buddhas and Wall Paintings in Bamiyan, Afghanistan. *Appl. Sci.* **2022**, *12*, 9476. [CrossRef]
62. Kuckova, S.; Hynek, R.; Kodicek, M. Identification of proteinaceous binders used in artworks by MALDI-TOF mass spectrometry. *Anal. Bioanal. Chem.* **2007**, *388*, 201–206. [CrossRef]
63. Tripković, T.; Charvy, C.; Alves, S.; Lolić, A.; Baošić, R.; Nikolić-Mandić, S.; Tabet, J. Electrospray ionization linear trap quadrupole Orbitrap in analysis of old tempera paintings: Application to nineteenth-century Orthodox icons. *Eur. J. Mass. Spectrom.* **2015**, *21*, 679–692. [CrossRef]
64. Sophie, D.; Monika, K.; Nicolas, G. Identification of animal glue species in artworks using proteomics: Application to a 18th century gilt sample. *Anal. Chem.* **2011**, *83*, 9431–9437.
65. Azémard, C.; Zazzo, A.; Marie, A.; Lepetz, S.; Debaine-Francfort, C.; Idriss, A.; Zirah, S. Animal fibre use in the Keriya valley (Xinjiang, China) during the Bronze and Iron Ages: A proteomic approach. *J. Archaeol. Sci.* **2019**, *110*, 104996. [CrossRef]
66. Yaroslav, D.; Timur, K.; Gavrilenko, L.; Solovyev, N. Targeted proteomics for the analysis of cultural heritage: Application of broadband collision-induced dissociation mass spectrometry. *Anal. Bioanal. Chem.* **2022**, *414*, 1723–1737.
67. Eva, E.; Karin, J.; Peter, P. Enzyme-linked immunosorbent assay. II. Quantitative assay of protein antigen, immunoglobulin G, by means of enzyme-labelled antigen and antibody-coated tubes. *BBA-Protein Struct.* **1971**, *251*, 427–434.
68. Pang, B.; Cheng, J.; Chen, Z. Comparision of Three Methods of Enzyme linked Immunosorbent Assay with A Monoclonal Antibody. *Chin. J. Biol. Control* **1999**, *1*, 32–35.
69. Duan, X.; Huang, X.; Huang, L.F.; Wei, H.; Lai, W.H. Detection of Listeria monocytogenes in Food by Sandwich ELISA. *Food Sci.* **2010**, *31*, 272–276.
70. Liu, J.Y.; Li, X.C.; Chen, D.K.; Zhang, Y.X.; Chen, J.; Wu, F.X. Establishment of indirect ELISA for detection of serum antibodies against canine parvovirus disease. *China AHS* **2006**, *3*, 27–29.
71. Liu, S.Z.; Feng, D.H.; Qian, C.F.; Sergei, A.E. The Residue of Carbaryl in Brassica chinensis and Apple Determined by Method of Antibody Immobilized Direct Competitive ELISA. *Chin. J. Pestic. Sci.* **2001**, *3*, 69–73.
72. Doreen, W.L.; Mohamed, S.; Abdel, R. Alterations in Rat Liver Microsomal Enzymes Following Exposure to Carbaryl and Malathion in Combination. *Arch. Environ. Contam. Toxicol.* **1985**, *14*, 451–457.
73. Bu, G.H.; Zheng, Z.; Zheng, H. Amethod for detecting β-lactoglobulin in cowmilk allergens by indirect competitive enzyme-linked immunosorbent assay. *J. China Agric. Univ.* **2008**, *6*, 71–76.
74. Cattaneo, C.; Gelsthorpe, K.; Phillips, P.; Sokol, R.J.; Smillie, D. Identification of ancient blood and tissue—ELISA and DNA analysis. *Antiquity* **1991**, *65*, 878–881. [CrossRef]
75. Cattaneo, C.; Gelsthorpe, K.; Phillips, P.; Sokol, R.J. Detection of blood proteins in ancient human bone using ELISA: A comparative study of the survival of IgG and albumin. *Int. J. Osteoarchaeol.* **1992**, *2*, 103–107. [CrossRef]
76. Schweitzer, M.; Hill, C.L.; Asara, J.M.; Lane, W.S.; Pincus, S.H. Identification of Immunoreactive Material in Mammoth Fossils. *J. Mol. Evol.* **2002**, *55*, 696–705. [CrossRef] [PubMed]
77. Lee, H.Y.; Atlasevich, N.; Granzotto, C.; Schultz, J.; Loike, J.; Arslanoglu, J. Development and application of an ELISA method for the analysis of protein-based binding media of artworks. *Anal. Methods* **2015**, *7*, 187–196. [CrossRef]
78. Palmieri, M.; Vagnini, M.; Pitzurra, L.; Rocchi, P.; Brunetti, B.G.; Sgamellotti, A.; Cartechini, L. Development of an analytical protocol for a fast, sensitive and specific protein recognition in paintings by enzyme-linked immunosorbent assay (ELISA). *Anal. Bioanal. Chem.* **2011**, *399*, 3011–3023. [CrossRef]
79. Hu, W.; Zhang, H.; Zhang, B. Identification of organic binders in ancient Chinese paintings by immunological techniques. *Microsc. Microanal.* **2015**, *21*, 1278–1287. [CrossRef]
80. Liu, Y. Micro-Trace-Identification of Ancient Proteinaceou Relics Based on Enzyme-Linked Immunosorbent Assay. Master's Thesis, Zhejiang Sci-Tec University, Hangzhou, China, 2017.
81. Wu, M.; Zhang, B.; Sun, G.; Jiang, L. Determination of lacquer contained in samples of cultural relics by enzyme-linked immunosorbent assay. *New J. Chem.* **2017**, *41*, 6226–6231. [CrossRef]
82. Xin, W.; Ming, Z.M.; Bing, J.Z. Detection of early Chinese organic-inorganic composite lime surface from the Lushanmao site, 4300 years ago. *J. Cult. Herit.* **2021**, *52*, 128–133.
83. Liu, L.; Shen, W.; Zhang, B.J. Determination of Proteinaceous Binders for Polychrome Relics of Xumi Mountain Grottoes by Using Enzyme-Linked Immunosorbent Assay and Immunofluorescence Microscopy. *Int. J. Conserv. Sci.* **2016**, *7*, 3–14.
84. Zhang, W. Development of Immunomagctic Bead Double Antibody Sandwich ELISA for the Detection of Trace Fibroin. Master's Thesis, China Jiliang University, Hangzhou, China, 2020.

85. Andrea, A.; Francesca, B.; Mariangela, P. Characterization of organic binders in a 13th century painted wooden panel: Comparison of ToF-SIMS and Dot-ELISA results. *Int. J. Mass. Spectrom.* **2018**, *430*, 63–68.
86. Zheng, H.; Yang, H.; Zhou, Y.; Li, T.; Ma, Q.; Wang, B.; Fang, Q.; Chen, H. Rapid Enrichment and Detection of Silk Residues from Tombs by Double-Antibody Sandwich ELISA Based on Immunomagnetic Beads. *Anal. Chem.* **2021**, *93*, 14440–14447. [CrossRef] [PubMed]

coatings

MDPI

Article

Consolidation of Fragile Oracle Bones Using Nano Calcium Sulfate Hemihydrate as a Protectant

Yan Liu [1,2,3], Ruicong Lu [1,2,3], Lu He [1,2,3], Ximan Wang [1,2,3], Lu Wang [1,2,3], Xinyan Lv [1,2,3], Kun Zhang [1,2,3] and Fuwei Yang [1,2,3,*]

1 China-Central Asia "the Belt and Road" Joint Laboratory on Human and Environment Research, Northwest University, Xi'an 710127, China; liuyan@nwu.edu.cn (Y.L.); 18066717745@163.com (R.L.); hl272103599@gmail.com (L.H.); wangximan996@163.com (X.W.); lwang810@foxmail.com (L.W.); 202120969@stumail.nwu.edu.cn (X.L.); 20199013@nwu.edu.cn (K.Z.)

2 Key Laboratory of Cultural Heritage Research and Conservation, Northwest University, Xi'an 710127, China

3 School of Culture Heritage, Northwest University, Xi'an, 710127, China

* Correspondence: yangfuwei@nwu.edu.cn

Abstract: Herein, a nano calcium sulfate hemihydrate suspension in an alcohol solvent was prepared and explored as a novel protectant for fragile oracle bones. The consolidation method involved first introducing the suspension and then adding water into the bones. Through this method, cohesive calcium sulfate dihydrate formed in the bones and can act as a reinforcing material. The protective effect was studied by scanning electron microscopy (SEM), energy dispersive X-ray spectroscopy (EDX), Fourier transform infrared spectroscopy (FTIR), X-ray diffractometry (XRD), hardness, porosity, and color difference determination. The results showed that such consolidation increased the strength of the bone samples significantly, and only slightly changed the appearance and porosity of the bone samples, indicating a good prospect for applying nano calcium sulfate hemihydrate in the conservation of indoor fragile bone relics.

Keywords: consolidating agent; bone relics; nano suspension; calcium sulfate hemihydrate

Citation: Liu, Y.; Lu, R.; He, L.; Wang, X.; Wang, L.; Lv, X.; Zhang, K.; Yang, F. Consolidation of Fragile Oracle Bones Using Nano Calcium Sulfate Hemihydrate as a Protectant. *Coatings* **2022**, *12*, 860. https://doi.org/10.3390/coatings12060860

Academic Editor: Gabriela Graziani

Received: 12 May 2022
Accepted: 16 June 2022
Published: 18 June 2022

1. Introduction

Oracle bones are animal bones that were used in the divination events of ancient China [1]. This type of bone divination dated back to as early as the late-Neolithic Age and flourished in the Shang Dynasty [2,3]. Inscriptions on the bones are believed to be the earliest Chinese characters and thus oracle bones are important for study on the early civilization of ancient China [4]. Unfortunately, the excavated oracle bones are often fragile as a result of the special processing procedure in divination and the long burial time of the bones [5]. Calcination is necessary in the preparation of oracle bones, the process in which the organic materials including collagen and fat in the bones are burnt out [6]. Additionally, some inorganic minerals such as calcium carbonate and carboxyapatite are partly dissolved during the long burial time [7]. The loss of the total organic matter and part of the inorganic minerals as described above leads to the decreased bulk density and mechanic strength of the oracle bones. Consolidation treatment on the bones is thus necessary for their preservation.

For a long time, organic materials such as paraffine, acrylic, and silicone resins have been the most widely used consolidating agents. They can work as glues and confer strength to weak bones once again. However, these organic consolidants are not sufficient in terms of weatherability and are apt to aging under photo damage, atmospheric oxidation, and microbial erosion [8]. After aging, their protective function decreases sharply, and negative consequences such as yellowing, embrittlement, and shrinkage occur, leading to the "preservation damage" of bone objects [9].

Due to their good weatherability, inorganic protective materials have been tested as alternatives. Calcium hydroxide was studied as a reinforcement material in archeological bones by Natali I. et al. [10]. After its introduction into the bone sample, it can react with carbon dioxide in the air and produce an adhesive aragonite. North A. et al. [11] studied the effect of ammonium hydrogen phosphate, which can react with calcite in the bone matrix and produce hydroxyapatite, thereby increasing the cohesiveness of the friable bones. Hydroxyapatite colloid was used directly as a consolidating agent of archeological ivories by Gong W. et al. [12] as it can be deposited on the surface of the ivories and form a strengthening layer. Recently, Liu Y. et al. [13] successively introduced a nano calcium hydroxide suspension and a solution of ammonium sulfate into bones, which could form calcium sulfate dihydrate and consolidate bone relics. Calcium sulfate dihydrate, also known as gypsum, has long been widely used in the repair of museum collections such as in pottery, porcelain, bones, and even metal wares [14–16]. It is stable enough and can hardly be aged in the museum environment. Calcium sulfate dihydrate can even be selectively cleared away by the scavenging agent of barium carbonate [17], making it a potential reversible protective material. However, the strategy of using calcium hydroxide and ammonium sulfate as the precursors of calcium sulfate dihydrate in a previous study [13] is challenging in practical applications. First, the distribution of calcium hydroxide and ammonium sulfate in the bone substrate is not uniform, which is also the case of the resulting gypsum product and its consolidation effect. Second, the irritant ammonia gas will be generated during the treatment process, which is an undesirable by-product during the reaction between calcium hydroxide and ammonium sulfate.

Hence, in this paper, a suspension of nano calcium sulfate hemihydrate in an alcohol solvent was prepared and explored as a novel consolidating agent for oracle bones under an indoor environment. The suspension liquid was introduced into the bone samples by surface permeability and then hydrated into an adhesive calcium sulfate dihydrate, which is illustrated in Figure 1. Due to the formation of the calcium sulfate dihydrate, the pores and cracks in the weathering bone samples were filled and the integrity of the samples was reestablished. The consolidation effect was evaluated by scanning electron microscopy (SEM), energy-dispersive X-ray spectroscopy (EDX), X-ray diffractometry (XRD), hardness, porosity, and color difference determination.

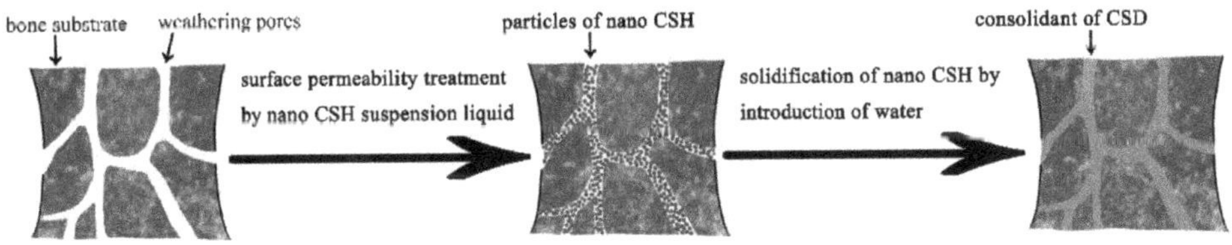

Figure 1. The schematic representation of the consolidation mechanism.

2. Experimental

Analytical grade reagents of calcium chloride ($CaCl_2$), 2-propanol (C_3H_8O), and sulfuric acid (H_2SO_4) were purchased from Sinopharm Group Co. Ltd., Shanghai, China. The suspension of nano calcium sulfate hemihydrate (CSH) in an alcohol solvent was prepared as follows. Anhydrous calcium chloride and sulfuric acid (90%) were first dissolved in ethylene glycol to yield a solution at concentration of 0.5 mol/L, respectively. Then, these two solutions were mixed equally to prepare the precipitate of CSH. After centrifugation and washing, the precipitate was re-dispersed in 2-propanol to produce a suspension with a concentration of 10 g/L.

The artificial oracle bones were prepared according to one of our previous studies [13]. The cow bones (1.0 cm × 1.0 cm × 2.0 cm) were calcined at 650 °C, soaked in a 4.5% solution of hydrochloric acid, washed with pure water, and finally dried naturally. Due to the loss of all of the organic collagens and part of the inorganic minerals, the artificially weathered bones are extremely similar to the archeological bones in terms of their chemical composition and physical properties.

The consolidation process of the bone samples is as follows. First, the suspension liquid of nano CSH was introduced into the bone samples by surface permeability. Then, the nano CSH was hydrated into calcium sulfate dihydrate (CSD) under a constant humidity of 85%. With the formation of CSD, additional cohesion was obtained and the weak bones were strengthened. The porosity and bulk density of the samples were determined by Picnometer Accupyc II 1240-Micromeritics (Centro di Geotecnologie, University of Siena, Siena, Italy). The mechanical strength of the samples was evaluated by a micro-hardness tester (HD-3000L, Shenzhen Wallok Testing Equipment Technology Co., Ltd., Shenzhen, China). The appearance change (ΔE) of the samples was evaluated by a reflectance spectrophotometer (WSC-S, D65 illuminant, 8°/d optical geometry, XiangYi Instrument (Xiangtan) Co., Ltd., Xiangtan City, China). The addition amount of nano CSH was 5–10% of the mass of the bone samples.

Scanning electron microscopy (SEM, FEI SIRION-100, 5.0 kV of accelerating voltage and 8.0 mm of working distance, Hillsboro, OR, USA) with energy-dispersive X-ray spectroscopy (EDX), Fourier transform infrared spectroscopy (FTIR, Nicolet 560, 4000–450 cm^{-1} range), and X-ray diffractometry (XRD, AXS D8 ADVANCE, Cu Ka radiation, scan range 2θ = 10–80°, Bruker, Billerica, MA, USA) were used to analyze the structure and composition of the samples. These analyses were performed on samples treated with 7% nano CSH.

The kinetic stability of the suspension liquid of nano CSH was evaluated by the absorbance measure at 300 nm using an ultraviolet/visible spectrophotometer (Varian Cary 100, SpectraLab Scientific Inc., Markham, ON, Canada).

3. Results and Discussion

The prepared calcium sulfate hemihydrate (CSH) suspension in 2-propanol is shown in Figure 2, which has a kind of milky liquid appearance. The micro-morphology and elemental composition of the CSH particles in the suspension are shown in Figure 3. The particles were stumpy and the size distribution roughly fits a Gaussian distribution with a range of 80–600 nm and the center distribution was about 200 nm (Figure 3b inset). This size scale is smaller compared to some of the previous studies, in which the particle sizes were usually bigger than 400 nm [18,19], and this smaller particle size is beneficial for good permeability. The particles are composed of O, S, and Ca elements from the EDX results. The phase composition of these particles was further confirmed by the XRD results in Figure 4. In the past, α-CSH and β-CSH were believed to be hardly distinguished by XRD spectra. However, according to recently published research [20], the peak position around 2θ = 30° shows a slight difference for the two kinds of CSH phases: the one for α-CSH is located at around 2θ = 29.80°, while this peak position for β-CSH is around 2θ = 29.70°. Therefore, the nano CSH particles presented here can be identified as the α type of the CSH from the aforementioned peak position located at 2θ = 29.81°. Additionally, the morphology of the CSH particles was also highly similar to that of the α-CSH observed by other researchers [21].

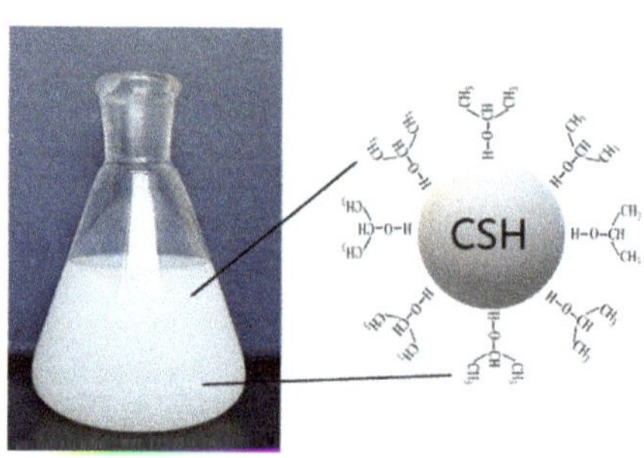

Figure 2. The CSH suspension in 2-propanol.

Figure 3. The SEM images of the CSH particles in the suspension liquid. (**a**) ×5000, (**b**) ×50,000.

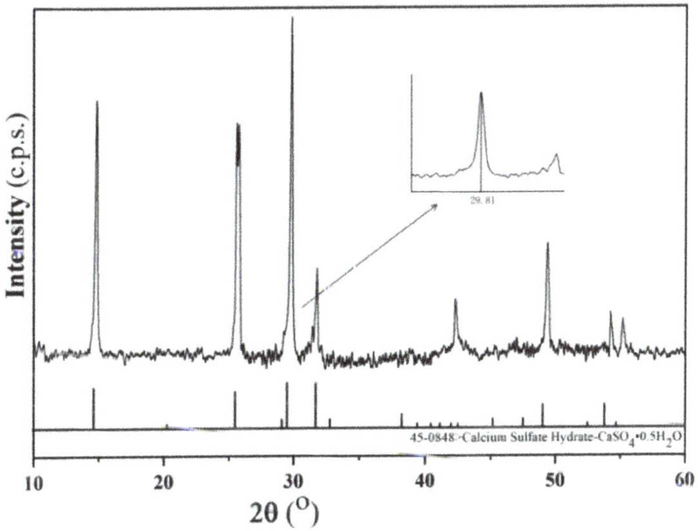

Figure 4. The XRD results of the CSH particles.

The suspension stability of CSH in 2-propanol was evaluated by absorbance measurements and the results are shown in Table 1. The absorbance was nearly constant during the test time, indicating a high kinetic stability of the suspension, which is due to both the small size of the CSH particles and the high steric hindrance of 2-propanol. As shown in Figure 2, 2-propanol absorbs on the surface of the CSH particles and forms an organic shield, which prevents agglomeration of the particles and setting of the suspension liquid [22]. This high stability of the nano materials in organic solvents has also been confirmed by previous studies, which is much better than in water (absorbance decreased to about 25% after 6 h) [23,24]. Stability is critical for a CSH suspension in practical applications, since it contributes to a deeper penetration of CSH into the open pores and cracks of the weathering bones to be consolidated.

Table 1. The kinetic stability of the CSH suspension.

Time (Hour)	0	1	2	3	4	5	6
Absorbance (%)	100	98	96.5	95	94.5	94.2	92

The morphology changes of the bone samples before and after treatment by the CSH suspension liquid are presented in Figure 5. As displayed in Figure 5a, the artificial oracle bone was coarse, porous, and loose due to the loss of the organic and inorganic components during the burning and subsequent burial process [25]. The suspension liquid of the nano CSH in 2-propanol was introduced into the bone sample by surface impregnation. After treatment, the open pores and fissures of the sample were patched up and the micro-structure seemed to be denser and flatter, as shown in Figure 5b. However, the particles of CSH were completely independent of each other at this time and can only serve as fillers. After the successive hydration treatment, further changes can be observed in Figure 5c,d. The separate CSH particles integrated together and a new board-like continuous structure was formed. This is mainly due to the dissolution–precipitation mechanism for the hydration of CSH [26]. As the hydration proceeds, the CSH is dissolved and then reorganizes to crystalline CSD [27]. In general, the typical morphology of the CSD is needle-like. However, its growth and morphology can be affected by the underlying matrix [10]. Here, we obtained a CSD with a board-like continuous structure, which may be related to the hydroxyapatite component of the bone samples.

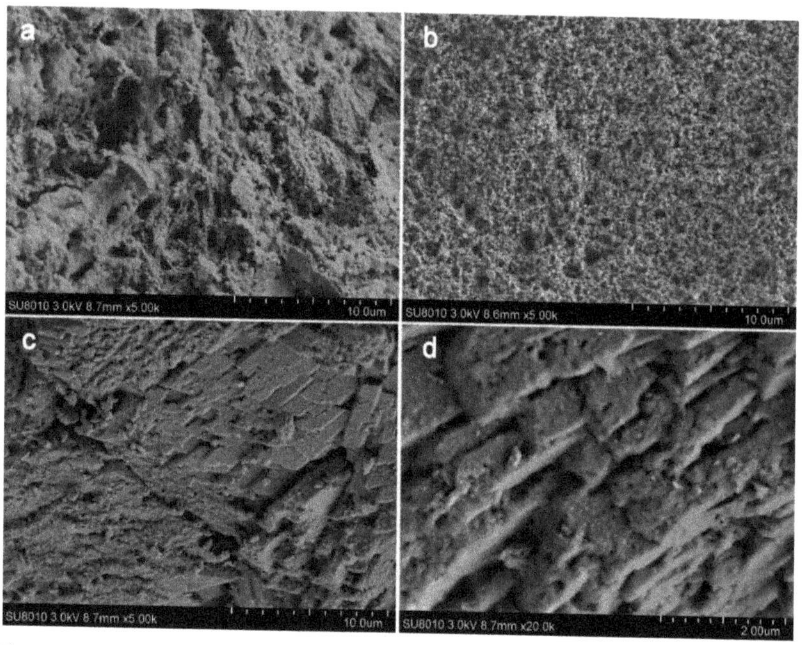

Figure 5. The micro-structure evolution of the bone samples. (**a**) Untreated, (**b**) treated by CSH, (**c**,**d**) treated by both CSH and water.

To investigate the possible mechanism of the morphology above, the composition of the bone samples was monitored during the consolidating process by FTIR and XRD analysis. The PO_4^{3-} bands (1090, 1018, 961, 606, 561 cm^{-1}) and the –OH bands (3555, 1633 cm^{-1}) in a in Figure 6 were from the bone sample, which was mainly composed of hydroxyapatite [28]. After introduction of nano CSH particles in the bone sample, the SO_4^{2-} bands (1141 and 666 cm^{-1}) and typical O–H band at 3612 cm^{-1} appear in b in Figure 6 and they are assigned to the calcium sulphate hemihydrate [26]. After further hydration treatment, the typical O–H band of calcium sulfate hemihydrate at 3612 cm^{-1} disappeared and the typical O–H bands of calcium sulfate dihydrate at 3405 and 1688 cm^{-1} appeared (c in Figure 6), indicating the transformation from calcium sulfate hemihydrate to calcium sulfate dihydrate [29]. From the results of the XRD in a in Figure 7, hydroxyapatite

(2θ = 25.9°, 28.9°, 31.7°, 34.1°, 39.8°, 46.7°, 49.5°, 53.2°) was the main component of the artificial weathering bone, which is consistent with that of archeological bones. After impregnation treatment, both calcium sulfate hemihydrate (2θ = 14.82°, 25.74°, 29.81°, and 31.86°) and hydroxyapatite were detected in b in Figure 7. The hydroxyapatite came from the bone sample itself. Calcium sulfate hemihydrate obviously comes from the suspension liquid of CSH, which is introduced during the impregnation treatment. After further hydration treatment, the diffraction peaks of calcium sulfate hemihydrate (2θ = 14.82°, 25.74°, 29.81° and 31.86°) disappeared and those for calcium sulfate dihydrate (2θ = 11.78°, 20.86°, 23.54°, and 29.26°) [30] appeared, as shown in c in Figure 7. Therefore, the specific appearance in Figure 5c is in fact the morphology of calcium sulfate dihydrate, which is the reaction product between calcium sulfate hemihydrate and water during the hydration procedure. Overall, the FTIR and the XRD results were in agreement with each other. Meanwhile, they also showed high consistency with the morphology changes displayed in Figure 5.

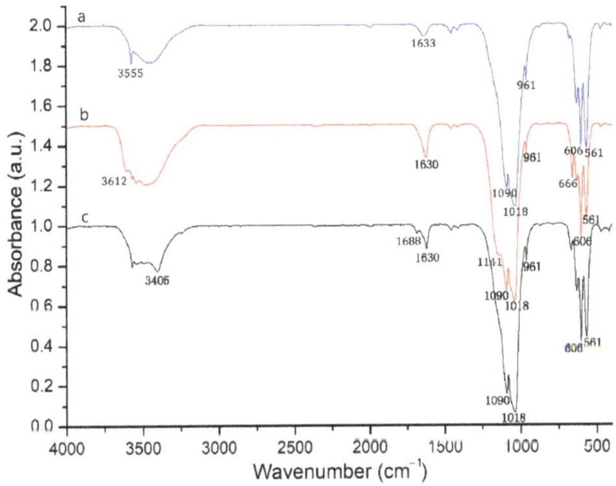

Figure 6. The FTIR results of the bone samples before and after the consolidation treatment. a. Untreated, b. treated by CSH suspension, c. treated by both the CSH suspension and water.

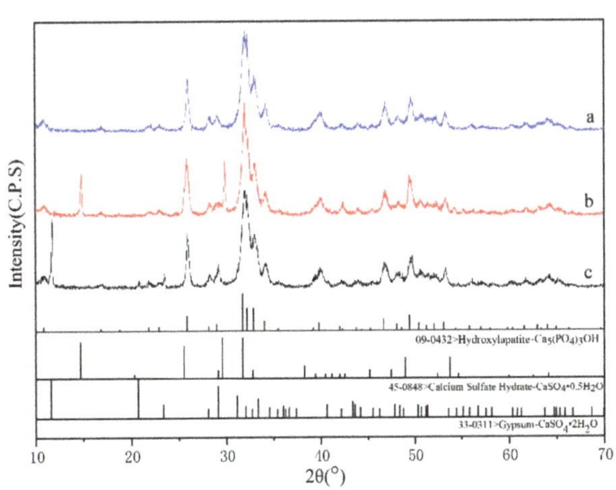

Figure 7. The XRD results of the bone samples before and after the consolidation treatment. a. Untreated, b. treated by CSH suspension, c. treated by both the CSH suspension and water.

The sectional structures of the bone samples were also observed by SEM and the results are shown in Figure 8. Before the consolidation treatment, the bone sample was rough and loose in appearance (Figure 8a). After the consolidation treatment, the bone sample became even and compact (Figure 8b). This means that the pores and cracks in the bone substrate had been filled by the calcium sulfate consolidant, which was proven further by the elemental distribution results of EDX. Sulfur was found throughout the cross section of the bone sample in Figure 8b (inset), showing a sufficient permeability of the CSH suspension. Similarly, the suspension of nano calcium hydroxide in the alcohol solvent also presented very good properties and was used in the conservation of the bone relics [10]. The small size of the nano inorganic consolidants plays an important role in these cases.

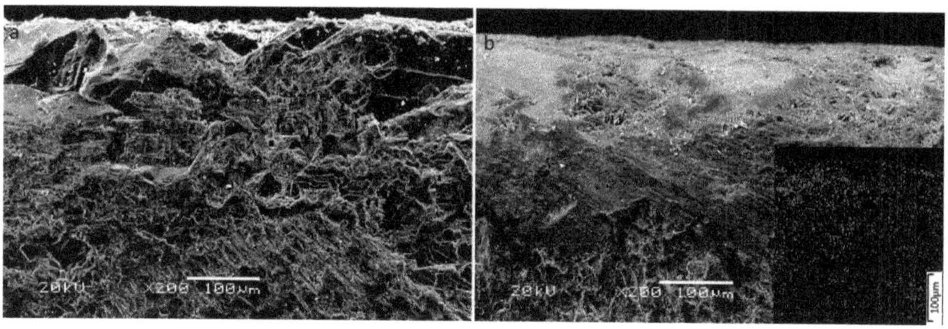

Figure 8. The cross–section of the bone samples. (**a**) Untreated, (**b**) consolidated by CSD.

The porosity, density, strength, and appearance changes were also investigated to assess the consolidation effect. After treatment, the porosity of the bone samples was reduced (Figure 9), while its density increased (Figure 10), confirming that the open pores and cracks of the bone samples were filled and the bone samples became denser. Generally, reduced porosity and increased density as a result of the reinforcement has been proven by many studies [31–33]. Surface hardness is adopted in the strength evaluation because of the inhomogeneity of the bone samples [34]. This index often refers to the deformation or damage resistance of the surface of an object. According to Figure 11, the hardness of the samples significantly increased after the consolidation treatment. This is obviously a result of the formation of CSD. After the hydration treatment, the particles of CSH were dissolved and furthered integrated together to form a continuous phase of the CSD phase [35]. Meanwhile, a strong connection was also established between the bone matrix and the produced CSD phase. Thus, the newly generated CSD phase can work as both a filling and reinforcing agent of weak bones. The change in appearance is one of the necessary evaluation indices for protective coating materials or methods used on cultural heritage [36]. It is commonly conducted by the determination of color difference. Generally, the threshold of color difference is 5.0 in the conservation of cultural heritage [37]. The color difference of the bone samples in the present study varied between 2.9 and 3.2 (Figure 12), which are all below 5.0, so are acceptable [38].

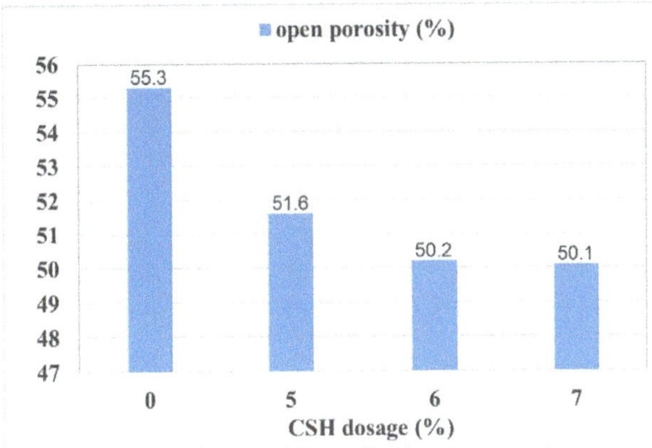

Figure 9. The open porosity change in bone with consolidation by CSH.

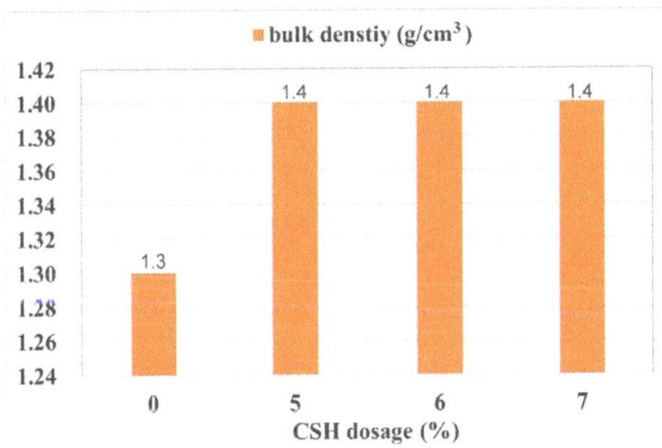

Figure 10. The bulk density change in the bone, samples after consolidation by CSH.

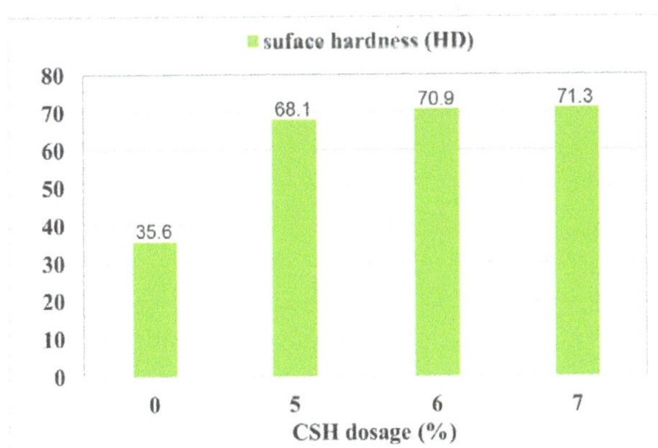

Figure 11. The surface hardness change in the bone samples after consolidation by CSH.

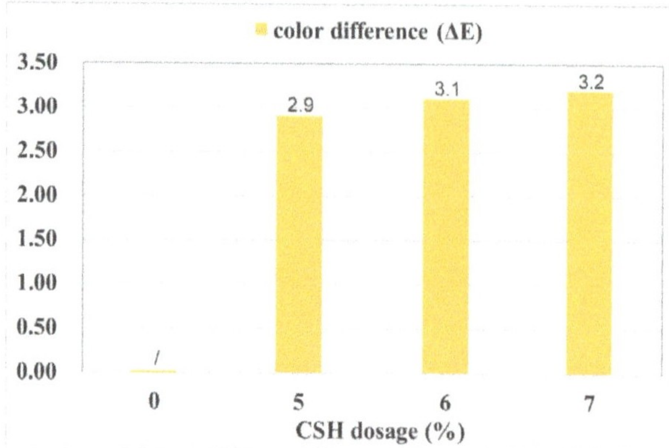

Figure 12. The color difference change in the bone samples after consolidation by CSH.

4. Conclusions

In this paper, nano CSH was successfully prepared by the reaction of calcium chloride and sulfuric acid in ethylene glycol solution. It was then re-dispersed in an alcohol solvent to obtain a suspension liquid with high kinetic stability and explored as a protective material for bone relics. In application, the suspension of CSH was first introduced into the bone samples, followed by the introduction of water, which hydrated the CSH into a continuous phase of CSD, filling the open pores, bridging the cracks, and providing additional strength for the weak bones. After the consolidation treatment, the porosity of the bone samples reduced, the strength of the bone samples increased, and the appearance change of the samples was acceptable. However, due to the low durability of gypsum in the open air, the proposed protective material is appropriate only for indoor porous weak bone objects and cannot be applied to any unsheltered or semi-sheltered objects.

Author Contributions: Conceptualization, F.Y.; Data curation, Y.L., R.L., L.H., X.W., L.W. and X.L.; Investigation, Y.L., R.L., L.H., L.W., X.L. and K.Z.; Methodology, F.Y.; Writing—original draft, Y.L.; Writing—review & editing, K.Z., R.L. and X.W. All authors have read and agreed to the published version of the manuscript.

Funding: This work was supported by the National Natural Science Foundation of China (B0501 21975202) and Key R & D Program in Shaanxi Province (2020SF-363).

Institutional Review Board Statement: Not applicable.

Informed Consent Statement: Not applicable.

Data Availability Statement: Data is contained within the article.

Conflicts of Interest: The authors declare no conflict of interest.

References

1. Liu, J.-H.; Ke, W.; Hwang, M.-C.; Chen, K.Y. Micro-Raman spectroscopy of Shang oracle bone inscriptions. *J. Archaeol. Sci. Rep.* **2021**, *37*, 102910. [CrossRef]
2. Demattè, P. The Origins of Chinese Writing: The Neolithic Evidence. *Camb. Archaeol. J.* **2010**, *20*, 211–228. [CrossRef]
3. Brunson, K.; Zhao, X.; He, N.; Dai, X.; Rodrigues, A.; Yang, D. New insights into the origins of oracle bone divination: Ancient DNA from Late Neolithic Chinese bovines. *J. Archaeol. Sci.* **2016**, *74*, 35–44. [CrossRef]
4. Anzhu, L. Oracle-Bone Inscriptions and Cultural Memory. *Front. Art Res.* **2020**, *2*, 63–73. [CrossRef]
5. Wang, K.; Hu, D.B. Hydroxyapatite: Collagen Biomimetic Composite Material in Conservation of Tortoise Shell Relics. *J. Natl. Mus. China* **2013**, *3*, 141–152.
6. Pérez, L.; Sanchis, A.; Hernández, C.M.; Galván, B.; Sala, R.; Mallol, C. Hearths and bones: An experimental study to explore temporality in archaeological contexts based on taphonomical changes in burnt bones. *J. Archaeol. Sci. Rep.* **2017**, *11*, 287–309. [CrossRef]

7. Kibblewhite, M.; Tóth, G.; Hermann, T. Predicting the preservation of cultural artefacts and buried materials in soil. *Sci. Total Environ.* **2015**, *529*, 249–263. [CrossRef]
8. López-Polín, L. Possible interferences of some conservation treatments with subsequent studies on fossil bones: A conservator's overview. *Quat. Int.* **2011**, *275*, 120–127. [CrossRef]
9. Han, X.; Huang, X.; Zhang, B. Morphological studies of menthol as a temporary consolidant for urgent conservation in archaeological field. *J. Cult. Herit.* **2016**, *18*, 271–278. [CrossRef]
10. Natali, I.; Tempesti, P.; Carretti, E.; Potenza, M.; Sansoni, S.; Baglioni, P.; Dei, L. Aragonite Crystals Grown on Bones by Reaction of CO_2 with Nanostructured $Ca(OH)_2$ in the Presence of Collagen. Implications in Archaeology and Paleontology. *Langmuir* **2014**, *30*, 660–668. [CrossRef]
11. North, A.; Balonis, M.; Kakoulli, I. Biomimetic hydroxyapatite as a new consolidating agent for archaeological bone. *Stud. Conserv.* **2016**, *61*, 146–161. [CrossRef]
12. Gong, W.; Yang, S.; Zheng, L.; Xiao, H.; Zheng, J.; Wu, B.; Zhou, Z. Consolidating effect of hydroxyapatite on the ancient ivories from Jinsha ruins site: Surface morphology and mechanical properties study. *J. Cult. Heritage* **2019**, *35*, 116–122. [CrossRef]
13. Liu, Y.; Hu, Q.; Zhang, K.; Yang, F.; Yang, L.; Wang, L. In-situ growth of calcium sulfate dihydrate as a consolidating material for the archaeological bones. *Mater. Lett.* **2021**, *282*, 128713. [CrossRef]
14. Milivojević, M. Excavation, reconstruction and conservation of steppe elephant from the clay pit of the building material factory "Toza Marković" at Kikinda (Serbia). *Bull. Nat. Hist. Mus.* **2011**, *4*, 51–64.
15. Ahmed, H.T. Restoration of historical artifacts and made available for exhibition in museums. *J. Am. Sci.* **2015**, *12*, 183–192.
16. Qin, Z.F. Restoration of a bronze grain receptacle with the pattern of Panhui (Spring and Autumn period). *J. Chinaese Antiq.* **2020**, *5*, 74–76.
17. Liu, Y.; Yang, F.; Wang, L. Exploratory research about the selective cleaning of calcium sulfate sediments on archaeological potteries. *New J. Chem.* **2020**, *44*, 7412–7416. [CrossRef]
18. Park, Y.B.; Mohan, K.; Al-Sanousi, A.; Almaghrabi, B.; Genco, R.J.; Swihart, M.T.; Dziak, R. Synthesis and characterization of nanocrystalline calcium sulfate for use in osseous regeneration. *Biomed. Mater.* **2011**, *6*, 055007. [CrossRef]
19. Hazra, C.; Bari, S.; Kundu, D.; Chaudhari, A.; Mishra, S.; Chatterjee, A. Ultrasound-assisted/biosurfactant-templated size-tunable synthesis of nano-calcium sulfate with controllable crystal morphology. *Ultrason. Sonochem.* **2014**, *21*, 1117–1131. [CrossRef]
20. Yin, S.; Yang, L. α or β?-hemihydrates transformed from dihydrate calcium sulfate in a salt-mediated glycerol–water solution. *J. Cryst. Growth* **2020**, *550*, 125885. [CrossRef]
21. Fu, H.; Jiang, G.; Wang, H.; Wu, Z.; Guan, B. Solution-Mediated Transformation Kinetics of Calcium Sulfate Dihydrate to α-Calcium Sulfate Hemihydrate in $CaCl_2$ Solutions at Elevated Temperature. *Ind. Eng. Chem. Res.* **2013**, *52*, 17134–17139. [CrossRef]
22. Leukel, S.; Panthöfer, M.; Mondeshki, M.; Schärtl, W.; Ruiz, S.P.; Tremel, W.; Schaertl, W. Calcium Sulfate Nanoparticles with Unusual Dispersibility in Organic Solvents for Transparent Film Processing. *Langmuir* **2018**, *34*, 7096–7105. [CrossRef] [PubMed]
23. Borsoi, G.; Lubelli, B.; van Hees, R.; Veiga, R.; Silva, A.S.; Colla, L.; Fedele, L.; Tomasin, P. Effect of solvent on nanolime transport within limestone: How to improve in-depth deposition. *Colloids Surf. A Physicochem. Eng. Asp.* **2016**, *497*, 171–181. [CrossRef]
24. Yang, F.; Zhang, B.; Liu, Y.; Wei, G.; Zhang, H.; Chen, W.; Xu, Z. Biomimic conservation of weathered calcareous stones by apatite. *New J. Chem.* **2011**, *35*, 887–892. [CrossRef]
25. Figueiredo, M.; Fernando, A.; Martins, G.; Freitas, J.; Judas, F. Effect of the calcination temperature on the composition and microstructure of hydroxyapatite derived from human and animal bone. *Ceram. Int.* **2010**, *36*, 2383–2393. [CrossRef]
26. Chen, X.; Wu, Q.; Gao, J.; Tang, Y. Hydration characteristics and mechanism analysis of β-calcium sulfate hemihydrate. *Constr. Build. Mater.* **2021**, *296*, 123714. [CrossRef]
27. Saha, A.; Lee, J.; Pancera, S.M.; Bräeu, M.F.; Kempter, A.; Tripathi, A.; Bose, A. New Insights into the Transformation of Calcium Sulfate Hemihydrate to Gypsum Using Time-Resolved Cryogenic Transmission Electron Microscopy. *Langmuir* **2012**, *28*, 11182–11187. [CrossRef]
28. Youness, R.A.; Taha, M.A.; Elhaes, H.; Ibrahim, M. Molecular modeling, FTIR spectral characterization and mechanical properties of carbonated-hydroxyapatite prepared by mechanochemical synthesis. *Mater. Chem. Phys.* **2017**, *190*, 209–218. [CrossRef]
29. Ding, X.; Wei, B.; Deng, M.; Chen, H.; Shan, Z. Effect of protein peptides with different molecular weights on the setting and hydration process of gypsum. *Constr. Build. Mater.* **2022**, *318*, 126185. [CrossRef]
30. Pan, Z.; Lou, Y.; Yang, G.; Ni, X.; Chen, M.; Xu, H.; Miao, X.; Liu, J.; Hu, C.; Huang, Q. Preparation of calcium sulfate dihydrate and calcium sulfate hemihydrate with controllable crystal morphology by using ethanol additive. *Ceram. Int.* **2013**, *39*, 5495–5502. [CrossRef]
31. Ibrahim, M.M.; Mohamed, W.S.; Mohamed, H.M. Evaluation of the efficacy of traditional and nano paraloid b72 for pottery consolidation. *Int. J. Conserv. Sci.* **2022**, *13*, 15–30.
32. Al-Dosari, M.A.; Darwish, S.; El-Hafez, M.A.; Elmarzugi, N.; Al-Mouallimi, N.; Mansour, S. Effects of Adding Nanosilica on Performance of Ethylsilicat (TEOS) as Consolidation and Protection Materials for Highly Porous Artistic Stone. *J. Mater. Sci. Eng. A* **2016**, *6*, 192–204. [CrossRef]
33. Terzu, R.; Baraj, E.; Förester, C.; Kropf, H.; Xhaxhiu, K.; Come, M. Ananalytical study of marble consolidation by oxalate precipitation using density, ftir and powder-xrd measurements. *J. Eng. Process. Manag.* **2017**, *8*, 21–26. [CrossRef]

34. Luo, C.; Liao, J.; Zhu, Z.; Wang, X.; Lin, X.; Huang, W. Analysis of Mechanical Properties and Mechanical Anisotropy in Canine Bone Tissues of Various Ages. *BioMed Res. Int.* **2019**, *2019*, 3503152. [CrossRef] [PubMed]
35. Van Driessche, A.E.S.; Benning, L.G.; Rodriguez-Blanco, J.D.; Ossorio, M.; Bots, P.; García-Ruiz, J.M. The Role and Implications of Bassanite as a Stable Precursor Phase to Gypsum Precipitation. *Science* **2012**, *336*, 69–72. [CrossRef] [PubMed]
36. Martínez, M.; Calero, A.I.; Valero, E.M. Colorimetric and spectral data analysis of consolidants used for preservation of medieval plasterwork. *J. Cult. Heritage* **2020**, *42*, 64–71. [CrossRef]
37. Ferreira Pinto, A.P.; Rodrigues, J.D. Impacts of consolidation procedures on colour and absorption kinetics of carbonate stones. *Stud. Conserv.* **2014**, *59*, 79–90. [CrossRef]
38. Concha-Lozano, N.; Lafon, D.; Sabiri, N.; Gaudon, P. Color thresholds for aesthetically compatible replacement of stones on monuments. *Color Res. Appl.* **2013**, *38*, 356–363. [CrossRef]

coatings

Article

Hybrid Tri-Cure Organo-Silicon Coatings for Monument Preservation

Cory B. Sims, Chamika U. Lenora and Joseph C. Furgal *

Department of Chemistry and Center for Photochemical Sciences, Bowling Green State University, Bowling Green, OH 43403, USA; cbsims@bgsu.edu (C.B.S.); lenora.chamika@gmail.com (C.U.L.)
* Correspondence: furgalj@bgsu.edu

Abstract: A coating system integrating three distinct chemistries was developed to protect materials used in monuments and construction. Initial curing is achieved using a UV-initiated thiol-ene reaction to form a non-impressionable/non-sticky surface. Second, amine/epoxy reactions form a firm surface adhesion and give mechanical strength through consolidation. Third, alkoxysilane sol-gel curing integrates the siloxane network while adding thermal stability, hydrophobicity, and a hardened surface. The final design utilizes a photoacid generator to increase the reaction speed of the second and third curing steps. The coating can be applied by spray, dip, or wipe on methods and exhibits a rapid non-impressionable surface (as fast as 10 min) that resists graffiti and environmental conditions, and is used and stored as a single-component system with a pot life exceeding six months. A series of experiments were used to determine the coating properties and durability, including field testing and accelerated weathering.

Keywords: tri-cure; coating; monument preservation; rapid cure; no mix; long pot life

Citation: Sims, C.B.; Lenora, C.U.; Furgal, J.C. Hybrid Tri-Cure Organo-Silicon Coatings for Monument Preservation. *Coatings* **2022**, *12*, 1098. https://doi.org/10.3390/coatings12081098

Academic Editor: Yulan Hu

Received: 28 June 2022
Accepted: 30 July 2022
Published: 2 August 2022

1. Introduction

Monuments are objects or places established for cultural or historical significance meant as reminders to future generations of lost individuals, outstanding accomplishments, or important events across the world. Commonly, these gravesites, statues, sculptures, buildings, and locations are constructed using various types of stones and metals to ensure the durability of the intended meaning. Many renowned ancient monuments such as Newgrange and the Great Sphinx of Giza have survived for thousands of years to the modern-day due to their stone construction; however, they have lost much of their original detail due to erosion of their sandstone and limestone components [1,2]. In the United States, many monuments are constructed to remember lost loved ones [3,4], events [5], honor great people [6,7], and even denote property boundaries [8]. The cost of these structures has a great range depending on the scale, from thousands of dollars for gravestones [3,4] to millions for national monuments [5–7], where private and/or public funding often pay for the cost.

With the expense required for installation, these entities must be adequately protected from environmental damage. There are three main erosion issues that monuments must persist through. First, acid rain can deteriorate the stone surfaces through dissolution (stone dissolved), or chemically alter surfaces through reactions with sulfur dioxide, forming discoloration and weakening the structure. Both of these processes are most detrimental to calcium-containing stones [9–11]. Oxidation is the second issue that affects metals, causing them to form metal oxides on the surface, weakening structural integrity, and causing discoloration, an example being The Statue of Liberty [9]. The third issue comes from fouling, when biological growth, such as lichens or mosses, produce acid on the stone's surface, causing local erosion [9,12,13]. When choosing how to protect monuments best, all three issues should be considered.

Historically, the preferred conservation method has been to use a series of surface coatings [9,10]. Many different systems have been adopted from use in other industrial applications due to their success (i.e., floor or automotive coatings) [14–16]. Organic polymers like epoxy resins, acrylates, and fluoropolymers have been used for this purpose due to their hydrophobic and protective capabilities, but they fall short in some areas when used individually [9,10,13,14,17,18]. While epoxies offer excellent surface penetration and mechanical strength, they are susceptible to photoinduced oxidation, discoloration, biological surface growth, are two-component systems, and have short working times [9,13,14]. Bisphenol A (dian) and cycloaliphatic epoxy systems are especially known to discolor and undergo oxidation over time [13]. Acrylics often offer surface reinforcement, are photostable, and may have anti-fouling capabilities, but they tend to be hard to remove and take a long time to cure [9,13,14]. Fluoropolymers provide a highly hydrophobic surface with photostability, but the hydrophobicity can make them difficult to handle, incompatible with many surfaces and environmentally unfavorable [14,18,19]. Despite this, there are numerous perfluoro compounds, both linear chains and cyclics such as perfluoropyridyls, which have increased the compatibility of these molecules with polymeric systems including siloxane networks [20,21]. The usage of these perfluoro groups have given R-alkoxysilane systems contact angles consistently measuring between 80–105° depending on the R-group chain length and branching [18,22].

Other coatings have used polysiloxanes for their photo and thermal stability and success in other industrial applications, but over time they degrade through hydrolysis [9,10,15,16,18,22–27]. Alkoxysilanes, methoxy and ethoxy, have been successfully employed in the consolidation of limestone and marble, forming surface bonding networks [28]. These networks have a low thermal conductivity and have been implemented to reduce thermal gains and stress on substrates by integration into infrared reflective coatings [28]. Sol-gel systems (i.e., R-alkoxysilanes) have also shown considerable success in preventing oxidative processes on metal surfaces [15,16,18,25]. One frequently used alkoxysilane system is tetraethoxysilane (TEOS), however the polymerization of this system is relatively slow and often requires the addition of a catalyst. Many acids and bases have been used to increase the reaction rates, but most achieve curing times exceeding 10 h [9,23]. These silicon-based systems are of particular interest as the components can be obtained through green chemical sources such as rice hull ash [29–31]. While each system has its benefits; the cons lead to difficulty in their application for long-term protection.

Some coatings have been designed to overcome the aforementioned limitations by integrating organics and inorganics into a hybrid system [14,22,27]. The use of organic functionalized alkoxysilanes to achieve traits of both systems, including epoxy and fluorocarbon chains, has been successful in some cases [14,18,20–22]. Siloxane modifiability has allowed for easier integration of specific capabilities, such as hydrophobicity or antifouling for coating systems to reduce biological growth [32,33]. Others have integrated alternative networking side chains to reduce the stress of shrinkage that occurs in both organic and inorganic polymers such as thiol and alkene or alkynes [17]. Many protective systems have integrated photo curing methods to increase reactivity while reducing the direct introduction of acids and bases to the substrate. To do this, photo initiators are used to induce polymerization of components upon exposure to a specific wavelength of light through radical formation or photo induced acid generation [34]. By relying on light to start the curing process, these systems may eliminate the need to store components separately. Nature has created several methods to add hydrophobicity and insulative properties to plants, including the use of waxy layers and microscopic texturing. The integration of surface roughness with synthetic coatings has shown promise in generating biomimetic materials with hydrophobic and oleophobic properties [35,36]. Alternatively, surface modification through spacing or nanoparticle additives with diameters under 0.1 microns has shown an increase in water contact angles through surface roughness, increasing overall hydrophobicity [37–40]. Integration of these various methods have created unique hybrid systems with a combination of their desired individual characteristics.

Ideal protective systems would offer a rapid cure, long-life, surface hardness, versatile application methods/conditions, exhibit hydrophobicity, and resist photodegradation and acid rain. This article focuses on the development and characterization of a tri-cure hybrid organo-silicon coating system for primary use on porous substrates and subsequent application on non-porous materials. Sol-gel processes are used to form an oxidative resistant, thermal, and photostable siloxane network backbone, with functionalized side chains that undergo alternative curing processes. This design integrates a thiol and vinyl photoinitiated cure (thiol-ene) to rapidly form a hard surface and relieve network stress on the overall system while adding extra photostability. Next, epoxy, amine, and fluorocarbon side chains increased surface adhesion and penetration, mechanical strength, and hydrophobicity (lower surface energy). The overall network exhibits favorable qualities of each coating type while reducing the drawbacks frequently seen in their separate usage.

2. Experimental Methods

2.1. Materials

2.1.1. Materials for Formulations and Testing

All chemicals were obtained through reputable commercial sources and used without further purification unless otherwise noted in the methods section. Hexane and sodium chloride were obtained from VWR International. Calcium chloride dihydrate, dichloromethane, glacial acetic acid, magnesium chloride, nitric acid, sodium hydroxide, sulfuric acid, and 1-Butanol were purchased through Fisher Chemicals (Hampton, NH, USA). 3-aminopropyltriethoxysilane, (3-glycidoxypropyl)trimethoxysilane, 3-mercaptopropyltrimethoxysilane, (tridecafluoro-1,1,2,2-tetrahydrooctyl)triethoxysilane, and vinyltriethoxysilane were obtained through Gelest Inc. Isopropyl alcohol and methanol were purchased through EMD Millipore Corporation. 1,3-divinyltetramethyldisiloxane, poly(ethylene glycol) 400, and tetrahydrofuran were obtained through Sigma-Aldrich and triethoxy(1H,1H,2H,2H-nonafluorohexyl)silane from AmBeed. Diphenyliodonium hexafluorophosphate and tetrabutylammonium fluoride, 1 M in THF, came from Acros Organics, and Omnirad 819 from IGM Resins USA, Inc. (Charlotte, NC, USA). 1,2-Bis(triethoxysilyl)ethane was obtained from Accela and methyltriethoxysilane through Alfa Aesar. Octamethylcyclotetrasiloxane and 2,4,6,8-tetramethyl, 2,4,6,8-tetravinylcyclotetrasiloxane were purchased from TCI. Ethanol (200 proof) was purchased from Pharmco. A Vivosun Handheld Garden, Pump Pressure Sprayer, was purchased through Amazon Inc. (Seattle, WA, USA). A 10 μm syringe from Hamilton Company (Franklin, MA, USA.) Great Value distilled white vinegar, 5%, was locally purchased from Walmart Stores. Mixed vinylsilsesquioxane cages (T_8, T_{10}, T_{12}) were synthesized according to Sulaiman et al. [41]. Siloxane depolymerization solution was made according to Rupasinghe et al. [42].

2.1.2. Substrate Materials

All materials were obtained and coated as is unless otherwise noted in the substrate preparation section. Marble tiles (shaped to ~8.5 cm × 10 cm rectangles, primarily calcite and dolomite, ~0.3%–2.0% porosity) were purchased from Home Depot Inc. (Atlanta, GA, USA) [43,44]. Granite monuments (~30 cm × 60 cm × 10 cm, primarily quartz and feldspar, 0.1%–2.0% porosity) were a gift from Detroit Memorial Park Cemetery [45,46]. Glass microscope slides (1.2 mm, 25.4 mm × 76.2 mm, silica) were obtained from United Scientific Supplies, Inc. (Libertyville, IL, USA) Bricks (9.5 cm × 21 cm × 7.5 cm, primarily silica alumina and iron) and wood blocks (untreated pine) were donated by Furgal Contracting Company, Warren, MI, USA. Steel samples and posts (non-galvanized steel) were donated by Nucor Marion Inc., Marion, OH, USA. Microflex powder-free nitrile gloves (XL, nitrile butadiene rubber) were obtained from Ansell Healthcare Products LLC (Iselin, NJ, USA).

2.2. Formulations

2.2.1. Substrate Preparation

Unless otherwise noted, samples described in the results and discussion used the substrates listed in Section 2.1.2. and were prepared as detailed below. To remove existing films or coatings, stone surfaces were stripped using 120 grit sandpaper (Ryobi 4″ × 36″ belt sander). Marble samples were shaped into ~8.5 cm × 10 cm rectangles during sanding and resurfaced after testing results were achieved for recoating/testing. Stone and glass samples were rinsed with distilled water, dried, sprayed with 0.1 M nitric acid, allowed to dry, then rinsed to imitate accelerated exposure to acid rain. Stainless steel surfaces were sandblasted to expose the raw surface and used as-is. Brick samples were scrubbed clean with a brush and water and then dried for several weeks. Wood blocks and nitrile gloves were used as received.

2.2.2. Synthesis of the Coating

General coating formulations were made as a one-pot system by volume percentage. Solids were calculated as a mass-volume percentage to yield a final formula. Components of each formulation were added to a 500 mL bottle and the resulting solution was then vortexed and used immediately or stored in a dark location at room temperature up to 1 year prior to use. For all applications, the bottle was shaken and vortexed immediately prior to usage. Ratios within the range depicted in Table 1 were used for all formulations, exceeding the best thirteen described in detail throughout this manuscript.

Table 1. Ranges explored and included in the final set of working formulations. Minimum and maximum volume percentages are given. Omnirad 819 and diphenyliodonium hexafluorophosphate are solids and were calculated by % m/V.

Reagent Volume Percentages	Min	Max
3-Glycidyloxypropyltrimethoxysilane	7	30
3-Aminopropyltriethoxysilane	6	30
Triethoxy(1H,1H,2H,2H-nonafluorohexyl) silane	0	12
(Tridecafluoro-1,1,2,2-tetrahydrooctyl)triethoxysilane	0	10
(3-Mercaptopropyl)-trimethoxysilane	0	18
Omnirad 819	0.2	1
Vinyltriethoxysilane	0	18
2,4,6,8-Tetramethyl, 2,4,6,8-tetravinylcyclotetrasiloxane	0	4
1,3-Divinyltetramethyldisiloxane	0	5
Vinyl Terminated Silsesquioxanes (mixed cages)	0	2
D4 Octamethylcyclotetrasiloxane	0	10
Methyltriethoxysilane	0	2
1,2-Bis(triethoxysilyl)ethane	0	7
Tetrabutylammonium fluoride, 1 M in THF	0	2
Diphenyliodonium hexafluorophosphate	0	8
H_2O	0	2
Poly(ethylene glycol) 400	0	2
Glacial Acetic Acid	0	2
Methanol	0	70
Ethanol	0	70
Isopropanol	0	70
1-Butanol	0	50

2.2.3. Coating Application

Indoor methods: Three application methods were used. Samples were sprayed until the surface appeared damp, dipped into the solution, and then allowed to dry while hung/suspended, or wiped on using tissue and the solution. Upon coating, samples were exposed to UV A light (18.7 mW, 200-Watt Mercury Arc Lamp with 320–390 nm filter, OmniCure Series 1500, Excelitas Technologies, Lumen Dynamics Group Inc., Mississauga, ON, Canada) for two minutes and allowed to sit for thirty minutes. For wood samples, each end was dipped into solution for 30 s twice, alternating ends, and then exposed to UV A for two minutes (total of four minutes). Lamp output measurements were obtained using a PM100D Energy Meter Console (Thorlabs, Newton, NJ, USA) equipped with an S121C standard photodiode power sensor (Si, 400–1100 nm, 500 mW, Thorlabs, Newton, NJ, USA).

Outdoor methods: Samples were spray-coated outside under sunny conditions with a UV index above 7 (>175 mW/m^2/s), temperature over 21 °C, and a clear forecast for 12 h after application. The application should be made in a back-and-forth sweeping motion like aerosol paint usage (SI-Gif S8). For small samples, disposable perfume bottles were used, applying approximately 1.2 mL of solution on an 8.5 cm × 10 cm marble surface (1 mL for 70.8 cm^2). For large samples, a handheld pressure sprayer was used applying the coating at an average rate of 4 mL/second at 20 pumps (40 mL/10 s).

2.3. *Analytical Methods*

Replication of tests was carried out where applicable using marble and glass substrates. Larger sample testing methods had limited replication due to the size and amount of substrate available. Limited sample sets utilized control surfaces to offer a comparison.

2.3.1. Attenuated Total Reflection Fourier Transform Infrared Spectroscopy (ATR-FTIR)

Attenuated total reflection (ATR) methods were used, and spectra were collected using OMNIC Spectra (Thermo Scientific, Waltham, MA, USA, 2017). Spectra were obtained on a Thermo Scientific ATR-FTIR (Nicolet iS5 Fourier Transform Infrared Spectrometer iD7 Attenuated Total Reflection, SN:ASB1817610). Liquid samples were placed on a ZnSe crystal, and the cured samples were obtained by coating a glass slide and placing the coated side down on the detector post-cure. The cured sample was obtained by coating a glass slide and placing the coated side down on a ZnSe crystal, and the liquid sample was placed directly on the ZnSe crystal and scanned. Time lapse FTIR was obtained by coating a glass slide and taking measurements in succession, where the time start at the initial exposure to UV light, and T 3 min was one min post exposure. Coating was scraped off monument samples and placed directly onto the ZnSe crystal for analysis. All samples were scanned from 4000 to 400 cm^{-1} for 16 scans with 0.121 cm^{-1} data spacing.

2.3.2. Thermal Gravimetric Analysis (TGA)

Thermal stabilities of samples were measured with a Hitachi STA7200 Thermal Analysis System (SN:19112035C1-01) using Software for NEXTA (Hitachi High-Tech Science Corporation, Tokyo, Japan, 2018, 2019). 3.11, 4.44, and 9.60 mg samples were placed into a ceramic crucible and heated at a rate of 10 °C per minute from 25 to 1000 °C and an airflow of 60 mL/min. Coating was applied to glass slides, removed by a razor post cure (2 days), and then dried in a vacuum oven at 45 °C for 24 h under 25 mmHg to remove any solvent prior to thermal stability testing. Material was then loaded into an Alumina pan and placed in the TGA for analysis.

2.3.3. Laser-Based Elemental Analysis

Elemental composition percentages were obtained using a Keyence VHX-7000 series Digital Microscope (Keyence Corporation of America, Broadview Heights, OH, USA) equipped with a Keyence Laser-based Elemental Analyzer EA-300 series. Microscope slides were coated, cured, and placed on the microscope stage. A class 1, 355 nm laser with 10 µm spot size was utilized for Laser-induced breakdown spectroscopy in an ambient environment.

2.3.4. Scanning Electron Microscopy (SEM)

Samples were prepared by mounting them to metal pegs with an adhesive strip, and then applying a conductive layer to the surface using a Hummer VI-A Sputter Coater (gold/palladium mix). Mounted samples were placed inside the instrument and images were obtained using a Hitachi S-2700 Scanning Electron Microscope (Hitachi, Tokyo, Japan). Series of images were obtained using 50–1000× magnification with working distances ranging from 13–25, 15–20 KV, and an aperture of 3 or 4. The coating was applied to different materials, and images of the coated and uncoated areas were taken using the same settings.

2.3.5. Pencil Hardness Assessment

Hardness tests were performed using an Elcometer 501 pencil hardness tester according to the manufacturer's instructions (7.5 Newtons, Elcometer Instruments Ltd., Manchester, UK). Coated glass and marble samples were assessed in increasing hardness. Results were confirmed by testing three different areas for each sample. Uncoated marble and glass samples were measured as standards.

2.3.6. Quantified Abrasion Resistance

Coated marble samples were weighed, sandpaper was placed on the surface with a weight (50.004 g, 1.43 kPa) on top, and the paper was passed across the surface (100 times). Samples were then blown clean (45 psi airflow) and reweighed. This was done in series using smooth (400 grit) and then rough (220 grit). Then the samples were sandblasted (30 s, 45 psi) and reweighed to calculate. The mass lost was calculated between each test and the observations made.

2.3.7. Coating Surface Analysis

Initial coating thickness measurements and topographic mapping was obtained using a Keyence VK-X1000 Series 3D Laser Confocal Microscope (Keyence Corporation of America, Broadview Heights, OH, USA). Coated marble and steel samples were used for direct imaging with the confocal microscope. Average coating thicknesses were obtained using an Alpha-Step IQ Surface Profiler (KLA Tencor, Milpitas, CA, USA). Glass slides were prepared using the dip-coat method, and five measurements were taken in opposite directions over the interface totaling ten thicknesses from which an average was obtained. Step height analysis was taken with a stylus force between 24.2–25.7 mg, 2000 μm scan length, 50 μm/s scan speed, 50 Hz sampling rate, 40 s scan time, 400 μm/23.8 pm sensor range, center bias adjustment, 1 μm resolution, contact speed of 5, and required radius of 5.0 μm.

2.3.8. Submersion Tests

One coated stone sample per formulation was submerged in various solutions for 24 h, 1 cycle, to investigate water and pH stability. Tested samples were exposed sequentially to distilled water, tap water, synthetic acid rain, and 0.1 M NaOH solutions were used with drying, and observations made in between exposures. Synthetic rain was made using 1 L of distilled water with a soda stream carbonated water system until a pH of 5 was achieved. Next, while stirring, five drops of nitric acid were added, and then sulfuric acid was added dropwise until a pH similar to acid rain (4.31) was achieved. After removal and drying, samples were inspected for delamination, cracking, and discoloration.

2.3.9. Chemical Resistance

Pieces of the cured coating were placed on a watch glass then submerged in various solvents for 24 h to check for reactivity and solubility. Tetrahydrofuran, hexane, dichloromethane, acetone, methanol, ethanol, isopropanol, 0.1 M nitric acid, 0.1 M sodium hydroxide, and water were used. Additional coating pieces were placed in a vial with the siloxane depolymerization solution [42] and allowed to stir in the solution for one month.

2.3.10. Graffiti Resistance

Two tests were conducted to investigate the effects of common vandalism tools on the final product. First, a line was drawn on samples using a Sharpie (Newell Brands). Then, tissue was used to wipe the permanent marker off the surface gently. Second, spray paint (Painters touch 2× Ultracover paint+primer, Rust-Oleum, Vernon Hills, IL, USA) was applied to the surface and allowed to dry. Then tissue was used to remove the paint. Sharpie testing was repeated twice on each sample analyzed for this method. Half-coated glass slides were used as a direct standard/sample comparison of the sharpie testing. Marble was not used as a standard as it absorbed the marker and paint is known to be difficult to remove from stone. No standard was used for spray paint.

2.3.11. Static Contact Angle Measurements

Water was added dropwise (~10 µL) via syringe to surfaces, and images were obtained using a Zeiss Stemi 2000-C light microscope (Carl Zeiss Meditec Inc., Dublin, CA, USA) with an AxioCam ERc5s camera. Contact angle analysis was performed using ImageJ software (1.53e, National Institutes of Health, Bethesda, MD, USA) and the manual points procedure through the contact angles plug-in. Values were derived by taking the mean of the right and left angles and reported with the standard deviation. Three measurements were made for each sample and most symmetrical images were used. Standard static contact angles were obtained for raw substrates which did not absorb the droplet.

2.3.12. Rust Resistance

A rusting solution was used to accelerate the occurrence of oxidation, and this was made by diluting 350 mL of white vinegar to 1 L with distilled water and adding 3.048 g NaCl.

Steel samples were dip-coated, and two testing methods were carried out. Two samples were hung and sprayed with the solution on both sides daily for one week, and one signpost sample was submerged in the solution for two weeks exposing the same length of coated and uncoated sample to the solution. Both methods consisted of half-coated samples which were used as a direct standard/sample comparison of the oxidative testing.

2.3.13. Hardwater Resistance

Imitation hard water was produced by adding 1.4729 g NaCl, 2.8143 g $CaCl_2 \cdot 2H_2O$, and 5.0883 g $MgCl_2$ to 250 mL of distilled water. This solution was applied dropwise to glass slides (half coated, dip-coat method) on both areas. Two heating methods were used. A heating gun was used to rapidly dry the solution and leave the salt residues, and then a clean tissue was used to wipe the residue off the surfaces. An oven was set to 45 °C, and the glass slide was baked for 24 h and once cooled, a tissue was used to wipe the residue off the surfaces. Both were observed for staining, smearing, and adhesion.

2.3.14. Environmental Stability Testing

Indoor methods: Marble samples were placed in an oven at 45 °C for 24 h, allowed to cool for 24 h, and then heated again for two cycles to investigate heat resistance. Samples were set in a freezer at −19 °C for 24 h then allowed to warm to room temperature; this was repeated after submerging samples in water for 24 h to imitate mixed rain and snow weather.

Outdoor methods: Samples were prepared outdoors, and the initial weather conditions [47] were recorded, including temperature, forecast, the chance of precipitation, wind, humidity, dewpoint, atmospheric pressure, UV index [48,49], air quality, ozone, nitrogen dioxide, sulfur dioxide, carbon monoxide, and particulate (µg/m^3). Weather data were recorded daily for the lifetime of the samples and will continue as long as they remain intact. Instances, where data was not recorded the day of, were noted, and historical records were used. Any unique phenomenon was noted, including regional wildfires, flooding, and nearby activities like construction which may alter the air quality of the test site.

2.3.15. Color Difference Analysis

Two images were selected of monument 1 from before coating and after 1 year had elapsed from application. These were auto white balanced in Adobe Photoshop 2022 (Version 23.1.1, Adobe Inc., San Jose, CA, USA) then the same region was magnified and selected on both to be analyzed for color details through the software. No other alterations were carried out on the selected images. The data was compared using the CIE76 [50] analytical method to produce an estimated change in color based on the CIELAB color space which corelates to alterations in the tint of an image. Values were assessed based on information in the literature.

3. Results and Discussion

The following sections will detail the rationale behind the coating development and potential uses, including curing processes, the optimization of formulations, coating procedures, characterization of formed coatings, and long-term stability testing. The thirteen formulas disclosed are provided as examples from the development process which led to the primary systems (10–13) which are discussed at a greater length below.

3.1. Curing Process and Design

The tri-cure system was developed using the three distinct chemistries shown in Scheme 1, which allow for the formation of a series of interdigitated macromonomers upon initiation. System composition was designed to integrate thiol-ene, epoxy/amine, and fluorocarbon chemistries (as needed) all on an alkoxysilane sol-gellable backbone. The thiol-ene reaction (macromonomer 1) is initiated by a photoreactive process to give rapid surface curing which achieved a consistent non-impressionable/sticky, yet still soft and relatively flowable surface within 30 min. To drive the initial thiol-ene cure, an Omnirad 819 photo initiator was used in the presence of either a UV-C mercury lamp (105 mW/cm^2) or sunlight for outdoor curing. This was initially optimized to occur within 10 min to prevent debris from sticking to the coating; however, often this was too fast to enable proper leveling and rearrangement in the coating. It was later found that a 30 min cure after 2 min of irradiation time was a more balanced time frame to impart leveling and reduce debris simultaneously.

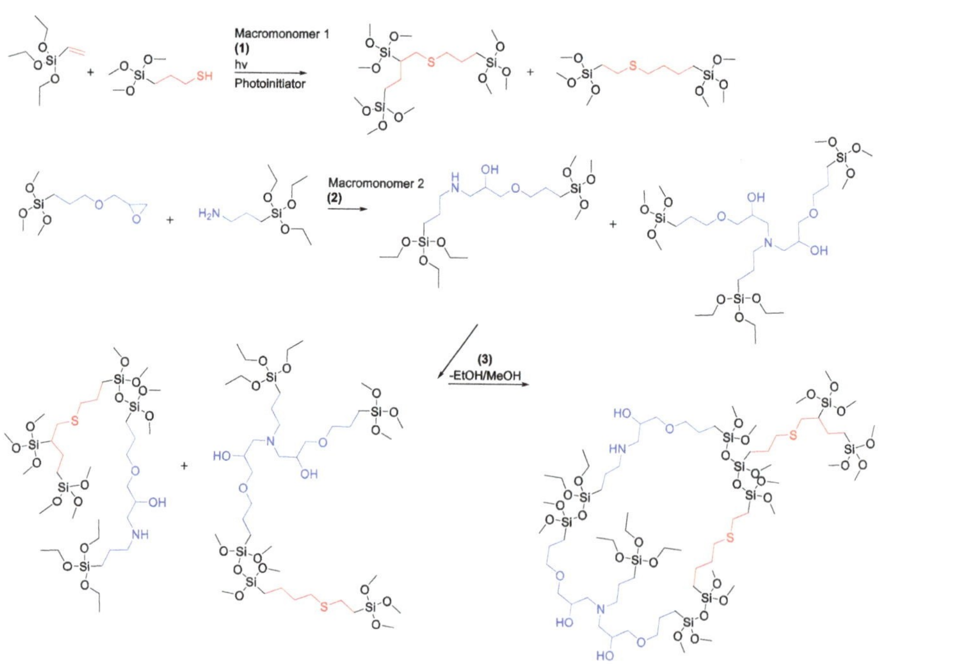

Scheme 1. General mechanisms of the three distinct chemistries in the system: (**1**) photoinitiated thiol-ene, (**2**) amine-epoxy, and (**3**) alkoxysilane polymerization in an example of the crosslinking process.

The amine-epoxy components are designed to provide further penetration into the surface, add surface adhesion and mechanical strength to the coating as the secondary curing process (macromonomer 2). Integrating alkoxysilanes with epoxy-amine systems has been shown to greatly reduce discoloring and photodegradation [13]. Initially we relied on catalyst free cure methods to drive the epoxy ring opening but found that it took more than a few days for the amine-epoxy reaction to take place. Therefore, we added a secondary photo-active catalyst to aid in acid-catalyzing this reaction and consequently the final sol-gel polymerization to tie macromonomers 1 and 2 together. A photoacid generator (PAG), diphenyliodonium hexafluorophosphate, was chosen as the catalyst since it can interact with the alcohols used as diluents in the system to form protons on demand. This also allowed for the same UV process to drive both the radical and acid cure processes together.

The third cure process took the two macromonomers formed in the first two cure steps and used the alkoxysilane components of those systems to couple everything together in a sol-gel reaction process. This procedure typically occurs more slowly over time (days to weeks) through environmental moisture cure processes and offers thermal and photostability while unifying the polymer network. The byproduct of the alkoxysilane reaction is a mixture of short chain alcohols, same as the solvents used, which hinders any premature siloxane formation prior to evaporation and exposure to moisture. In later formulations, the addition of a photoacid generator accelerated both the amine-epoxy and the alkoxysilane curing processes to under several hours, much faster than most traditional alkoxysilane coatings [23]. Other R-group functionalities on alkoxysilanes can also be tied in here (i.e., perfluoroalkylalkoxysilanes) through the alkoxysilane cross linkage. Fluorocarbon components were used in low ratios to increase the hydrophobicity and due to their chemical and biological resistance. Amounts used with all fluorocarbon species yielded similar contact angles (95–105°) as those seen in 1H,1H,2H,2H-perfluorooctyl trichlorosilane when used in an alkoxysilane system [22]. The combination of the above components was investigated through various chemical and physical properties analysis as detailed further below.

3.2. Chemical Characterization

Chemical analysis was carried out using ATR-FTIR and TGA to probe the curing. A representative ATR-FTIR was taken for the solution and spray coated glass surfaces of formula 10 to observe changes occurring during the curing process at time = 0 and after 1 week (Figure 1). The solution spectra primarily show the presence and loss of methanol and isopropanol by the O–H stretching at 3300 cm^{-1}, C–H peaks at 2900 cm^{-1}, C–O stretching at 1085 cm^{-1}, and C–O stretching 1033 cm^{-1}, and O–H Bending at 950 cm^{-1} which are mostly absent from the cured sample. Additionally, the absence of 1° amine and alcohol peaks in the final product suggests the reactivity of the secondary epoxy and amine process occurred; however partial products may still be present in the system. S–H peaks are absent around 2600 cm^{-1} or C=C–H peaks around 3000 cm^{-1} in the coated spectra, suggesting minimal amounts of vinyl remain in the system, supporting that the thiol-ene reaction reaches completeness. Furthermore, the peaks at 1012 and 875 cm^{-1} in the cured spectra indicate the presence of a combination consisting of Si–O, C–O, and Si–O–C bonds which would be expected of alkoxysilanes and partially reacted species, supporting the occurrence of the tertiary sol-gel reaction expected to take place over a more extended time period.

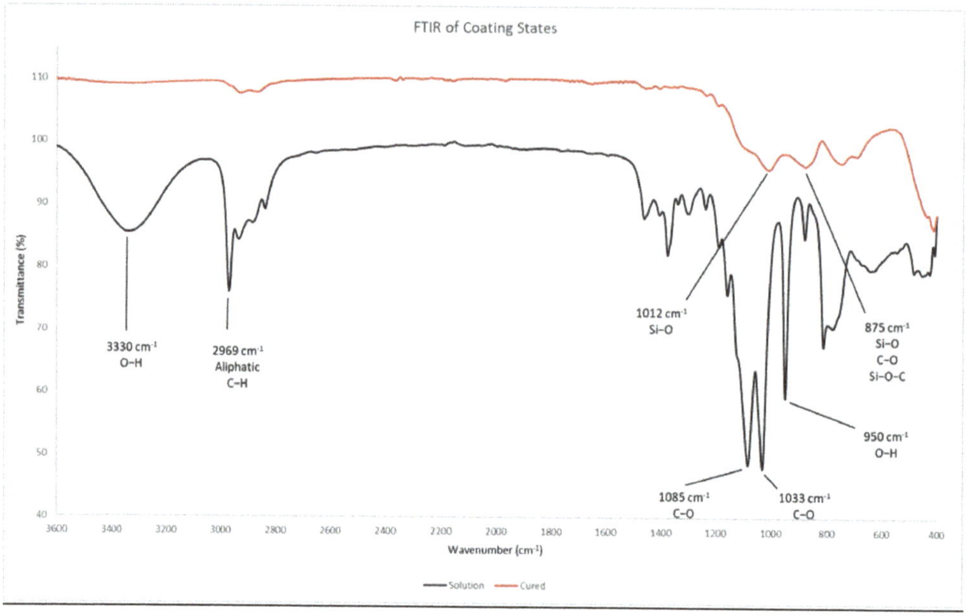

Figure 1. ATR-FTIR of the coating solution (black, t = 0) and cured coating on glass (red t = 1 week)) of formula 10.

These findings are supported through a timelapse ATR-FTIR of formula 11, from 1 min post cure (T_3) until the surface was hard to the touch (T_{35}). Within the first 5 min, there is a loss of the O–H stretching at 3400 cm^{-1}; and over time the Si–O band at ~1100 cm^{-1} gets stronger, suggesting the formation of Si–O–Si bonds (Figure S1). After 209 days passed, peaks were still observed around 3300 cm^{-1}, 2930 cm^{-1}, and 1010 cm^{-1} with notable peaks at 1590 cm^{-1} and 1410 cm^{-1} suggesting the presence of amine and O–H groups in a sample of formula 11 (Figure S2). Once monument 1 (formula 11) reached 365 days of field testing, a small section of the head stone coating was removed by a razor blade and analyzed with ATR-FTIR (Figure S3). This spectrum shows similar peaks as those seen previously, however the indication of an amine peak at 1590 has shifted to 1650 cm^{-1}. FTIR of formula 12 at 1 day and 178 days after application on a glass slide, and of a 282-day old outdoor sample all exhibit similar peaks as those mentioned above; however, an additional amine peak is observed around 1650 cm^{-1} in these samples (Figures S4–S6). This suggest the presence of different amines as indicated to be likely in Scheme 1. Lastly, spectra were obtained for formula 13 (Figure S7) which shows the presence of thiol at 2600 cm^{-1} in addition to the peaks seen above.

Siloxane formation is further supported through TGA analysis (Figures S8–S13). When allowed to fully cure prior to analysis (2 weeks), these coatings remain stable with a $T_{d5\%}$ of 308 °C, 275 °C, and 262 °C for formulas 10, 11, and 12, respectively, which is within the anticipated range for siloxane network stability. After time elapsed some samples exhibited slightly lower stabilities with shifts in their $Td_{5\%}$ values to 254 °C (formula 11 at 209 days), 263 °C (formula 12 at 178 days), and 230 °C (formula 12 with 282 days of environmental exposure). All three samples show an indication of moisture by a slow decline in mass up to 100 °C with the outdoor sample showing the largest decrease in this region at 2% of the sample, and a $T_{d7\%}$ of 260 °C supporting this theory. Each of these samples indicate proper formation of Si–O–Si bonds and relatively constant thermal stabilities over time.

Elemental weight percentages were obtained using a Keyence Laser-based Elemental Analyzer EA-300 series. The laser-induced break down spectroscopy is performed under ambient atmosphere and can't measure nitrogen, preventing confirmation of amine presence. Despite this the analysis gives reasonable weight percentages for formulas 11, 12, and 13 when considering the relative amounts of elements in each system without solvent and the small sample area (10 μm diameter) (Table S1).

3.3. Formulation

Formulations were developed with several end-of-life goals in mind: (1) The solution is made in a single container as a no-mix system, (2) the final formulations have a pot life exceeding one year, (3) the solution can be applied as a spray-on coating, and (4) they have potential recyclability. Each of these goals have been achieved to different extents with pot lives of some coatings even achieving a year or more of dark stability.

Over 100 different formulations were developed, and all designs fell within the ranges of the components given in Table 1; however, one series resulted in a desirable set of model systems which will be addressed as formulas 1–13. These solutions are named in the order of their development and were used for the analysis of the properties detailed in this article (Table 2). Formulations were designed with rapid curing in mind and adjustments to how quickly the coating dries can be made by using longer chain alcohols for the carrier solvent, lengthening the time it takes for the network to settle/arrange on the surface. The developed systems may be applied through spraying, dipping, or wiping, but they were intentionally designed with pressurized pump-spray systems in mind for outdoor use. The intended use for monument preservation led to most of the testing being focused on sprayed-and-dipped applications utilizing marble/granite samples.

Table 2. The thirteen coating formulas resulting in the development of the final systems. All amounts are given as v/V%'s with the exception of Omnirad 819 and diphenyliodonium hexafluorophosphate which are given as m/V%'s.

Reagent %/Formula No.	1	2	3	4	5	6	7	8	9	10	11	12	13
3-Glycidyloxypropyl-trimethoxysilane	28.0	13.8	13.8	13.8	13.7	14.1	12.7	13.3	13.7	13.7	8.2	13.7	13.6
3 Aminopropyl-triethoxysilane	28.0	12.4	12.3	12.3	12.3	12.7	11.3	11.9	12.3	12.3	7.5	12.3	12.3
Triethoxy(1H,1H,2H,2H-nonafluorohexyl) silane	5.5	2.8	2.7	2.7	2.7	-	10.5	2.6	2.7	2.7	1.6	-	-
(Tridecafluoro-1,1,2,2-tetrahydrooctyl)tri-ethoxysilane	-	-	-	-	-	-	-	-	-	-	-	2.8	2.8
(3-Mercaptopropyl)-trimethoxysilane	9.0	4.8	4.9	4.9	4.9	5.0	4.5	4.7	4.9	4.9	2.9	5.0	5.0
Vinyltriethoxysilane	9.0	5.6	5.6	5.6	5.6	5.7	5.1	5.4	5.6	5.6	3.3	5.6	5.5
Omnirad 819	0.5	0.3	0.6	0.4	0.7	0.7	0.5	0.5	0.5	0.6	0.3	0.5	0.8
D4 Octamethylcyclo-tetrasiloxane	8.0	-	-	-	-	-	-	-	-	-	-	-	-
Diphenyliodonium hexafluorophosphate	-	-	-	-	-	-	-	3.4	0.1	0.2	0.1	0.2	0.3
Methanol	12.0	60.3	60.2	60.3	-	61.8	55.4	58.2	60.1	-	36.1	18.1	18.1
Ethanol	-	-	-	-	60.1	-	-	-	-	-	-	-	-
Isopropanol	-	-	-	-	-	-	-	-	-	60.1	40.1	41.8	-
1-Butanol	-	-	-	-	-	-	-	-	-	-	-	-	41.7

Initial design began with a series of solutions containing various ratios of components with relatively low solvent percentages (10%–15%) and octamethylcyclotetrasiloxane (D4) as a non-reactive diluent to improve surface leveling. These were designed to investigate the effects of altering the ratios of components pertaining to the different curing methods. This was achieved by systematically increasing one set of components while decreasing another over a range of volume percentages and repeating the process for multiple combinations of adjustments (i.e., increasing thiol-ene/lowering adhesion, increasing adhesion/decreasing fluorocarbon). The resulting coatings were inspected for their hydrophobicity; systems which did not delaminate after initial curing were chosen for further exploration, including formula 1.

The next set of solutions were designed to investigate adjustments using formula 1 as a basis. From these samples it was decided to remove D4 as the system was intended for spray application and this component added unnecessary cost and health risk to the end-user. While adhesion was successful in these D4-free systems, delamination was still occurring during the secondary and tertiary curing processes days later, and thus a series of solutions were designed to see the effects of increasing the solvent ratio from <15% to 25%, 50%, and 70%. From these dilutions the delamination occurring upon hardening was resolved, suggesting earlier systems were too thick. These thinner samples were coated on glass to offer an even surface to compare static contact angles (see physical characterization). Formula 2 was found to have the highest contact angle from the diluted solutions when inspected on glass and became the new base formulation to build on.

When designing the various formulations, increased hydrophobicity and ensured long-term survival to water exposure were tested. Included in these tests were the implantation of different vinyl-containing compounds as crosslinkers, the potential for an increased reaction rate through the addition of acetic acid to the solution, as well as modifications to the surface by way of nanoparticle formations [38], both of which were unsuccessful. Then the best samples were chosen from all previous formulations based on contact angles, adhesion, and removability of permanent marker ink. Of these variations, it was deemed an unnecessary expense to implement nanoparticles, addition of acetic acid lowered the contact angle slightly, and most alternative crosslinkers delaminated. From these results, formula 3 (Table 2) was chosen for further testing due to its optimal performance. However, delamination was still observed in this system during the long-term exposure testing (see physical characterization) where it was submersed in water for 24 h.

To resolve the delamination issues, several sets of this formula were investigated with varying coating thicknesses, UV exposure times, alternative solvents, annealing, and the addition of tetrabutylammonium fluoride (TBAF) as a co-catalyst aid in the sol-gel process. From these tests it was determined that coating thicknesses (~ 2–10 μm) resulting from applying approximately 1.2 mL of solution on an 8.5 cm × 10 cm marble surface (1 mL for 70.8 cm^2) eliminated delamination from curing and submersion testing. The new application parameters resulted in formula 4, where the ratios remained relatively the same as formula 3. Samples which implemented TBAF for the sol-gel process had delamination and flaking issues upon curing, and thus were not pursued. Additionally, these tests showed that increasing the cure time and annealing at 45 °C for 24 h resulted in an increase in static contact angles over formula 3 which had 2 min of UV exposure and cured at room temperature. While increasing hydrophobicity/contact angles was a main priority, annealing and elongated cure times were not used during batch sampling indoors, but showed promise for future outdoor applications. Ethanol was used as the carrier solvent for formula 5 and, while it had similar graffiti resistance and contact angles, it delaminated after water submersion testing where system 4 survived water, acid rain, and base solution submersion testing and both utilized similar application parameters.

After water sourced delamination was resolved through the success seen in formula 4, development focused on the effects of coating method and surface preparation to enhance overall adhesion. Results suggested that application method (spray, wipe, or dip coated) had little alteration on the resulting coating, however when comparing stone pretreated with water, vinegar (5%), or 0.11 M Nitric Acid, an improved contact angle was seen in the latter. The collective findings at this point suggested the use of a thinner coating is necessary overall and nitric acid prepped surfaces (mimicking acid rain exposure) are ideal for indoor stone applications.

Further investigation looked at the additive effect of the fluorocarbon species through formulas 6 and 7 which confirmed its role in hydrophobicity. Minimal amounts of triethoxy(1H,1H,2H,2H-nonafluorohexyl) silane were used due to the purchase cost, but removing it completely resulted in a decreased contact angle of 92.6° in formula 6. Testing showed that increasing the amount used resulted in an increase from 97.1° (~3%) to 101.5° (~10%) in the observed contact angle between formulas 4 and 7, however, to explore ratios higher than this range was deemed an unnecessary cost. Similarly, a set of samples were made without the thiol-ene components according to the same conditions. The same samples delaminated, suggesting the thiol most likely relieves tension in the alkoxysilane network.

In an effort to increase the overall hydrophobicity and reaction rate, a photo acid generator (diphenyliodonium hexafluorophosphate) was added in small amounts and the effects on contact angle were compared through formulas 8 and 9. When used in high concentrations, as with formula 8, the system was non-impressionable within 9 min and lower amounts maintained the previously achieved 30-min surface hardness. Ultimately, small amounts of the photo acid generator (PAG) were utilized, as adding too much decreased the observed contact angle, where a higher concentration in formula 8, and lower amount in formula 9 showed a sizeable difference (95.9° vs. 101.5°). These systems were used in the initial outdoor testing, but the solvent in the first samples of each evaporated before curing could be achieved. In response, the formulas were diluted with isopropanol where a diluted formula 8 is still in the field on a smaller marble sample at 371 days, as of this writing. Variations of formula 9 were made using isopropanol as the sole solvent (formula 10) and mixed with methanol (formula 11). Both systems were used for various properties testing to explore the effects of altering the solvent.

The mixed solution 11 was used in the first upscale application on a granite headstone (monument 1), which remains unaffected after 371 days of outdoor exposure at this time. Two more systems were designed which used an alternative fluorocarbon, (tridecafluoro-1,1,2,2-tetrahydrooctyl) triethoxysilane, which is longer and cheaper. The first (system 12) uses the isopropanol dilution and has comparative characteristics to formula 11. Alternatively, formula 13 was designed to lengthen the working time of the solution and used a dilution of butanol to increase the time it takes to achieve non-impressionability from 30 min, with the aforementioned systems, to between 68 and 97 min depending upon the application method and thickness. Collectively, formulae 6–13 have shown success in application and testing, while the ideal systems for monument preservation are formulas 11–13 depending on the user's needs for cost and working time. Formulas 11 and 12 were utilized for the majority of the detailed characterization and properties testing described below due to their initial results and large batch quantities since many tests required large amounts of coating to perform (i.e., dip coating for oxidation testing).

3.4. Application and Substrates

A series of substrates were compared for these coatings including marble, granite, brick, glass, steel, wood, and nitrile to demonstrate versatility, even though our initial goal was for the field of monument preservation. Table 3 gives a list of coating formulations and their applicable substrates with application methods. As discussed above three primary application methods were used depending on substrate scope, including spray, wipe, or dip coating. All 3 methods were successfully used for the marble and glass samples, whereas dip coating was preferred for the steel or wood, and wipe-on for the nitrile samples. Further analysis of these application methods and their performance in representative samples is discussed throughout the manuscript for the different types of samples. Most of the comprehensive testing was conducted using formulas 11–13 which represent the most diverse substrate sampling.

Table 3. Substrates and applications investigated for each of the formulas used. Each material is noted what application methods were used in parenthesis as sprayed (S), dipped (D), and wiped (W). Not all samples or methods listed were used in analysis or measurements.

Formula No.	Application Substrates and Methods						
1	Marble (S)	Glass (S)	-	-	-	-	-
2	Marble (S)	Glass (S)	-	-	-	-	-
3	Marble (S)	-	-	-	-	-	-
4	Marble (S)	-	-	-	-	-	-
5	Marble (S)	-	-	-	-	-	-
6	Marble (S)	Glass (S)	-	-	-	-	-
7	Marble (S)	Glass (S)	-	-	-	-	-
8	Marble (S)	Glass (S)	-	-	-	-	-
9	Marble (S)	Glass (S)	-	-	-	-	-
10	Marble (S)	Glass (S)	-	-	-	-	-
11	Marble (S)	Glass (S, D, W)	Granite (S)	Brick (S)	Nitrile (S, W)	Steel (S)	-
12	Marble (S)	Glass (S, D, W)	Granite (S)	Brick (S)	Nitrile (S, W)	Steel (S, D, W)	Wood (D)
13	-	Glass (S, D)	-	-	-	-	-

3.5. Physical Characterization

Surface inspection was performed through SEM images to investigate how the coating adheres to the various substrates and its general appearance/coverage. Initial pictures focused on marble samples coated with formula 11. In both the 50× and 300× images, there is a drastic difference in the surface once the solution settled into the grooves of the calcite (marble), giving a much smoother appearance (Figure 2). Coated regions have a lower resolution since the electron beam interacts with the transparent coating, resulting in a dark, semi-opaque appearance of features which becomes more intense the thicker the coating is. There is evidence of micron-sized protrusions in the 1000×, coated image, suggesting a thin, secondary application may be necessary to protect the rough stone. The abrasive removal of the top layer of stone using sandpaper is evident through the tracts seen in Figure 2C, and the transition from the raw to the protected surface is easily distinguished. Due to air pockets trapped under the coating, imaging a thicker surface resulted in fissures forming in these areas upon exposure to high vacuum, which is seen in all porous or rough surfaces imaged and is verified through imaging of a smooth substrate detailed later.

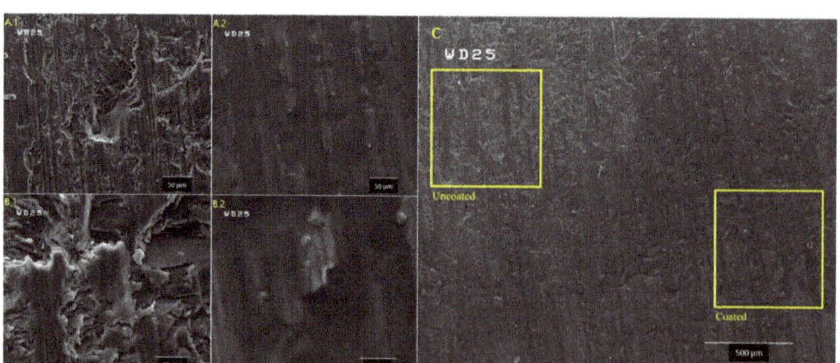

Figure 2. SEM images at 300× (**A**) and 1000× (**B**) magnifications comparing the bare (1) and coated (2) surfaces of a marble sample, primarily calcite, treated with formula 11. SEM image (**C**) of the transitional region between surfaces at 50× magnification, the darker region has been coated.

Later in the development process, a piece of the coated monument 1 (formula 11) was chiseled out of lettering after 331 days, the extrusion inside of the R in "FORCE" (detailed below and shown in Figures S28 and S29). This was used to take SEM images of the same coating after long-term weathering to confirm the presence of and stability of the coating on granite (quartz and feldspar) surfaces. As seen in Figure S14, the coating has formed a complete layer notable by the observed smoothness and continuous color which includes confirmed regions. Like what is observed in the other stones coated with formula 11, areas where the coating formed an air pocket due to the porous nature of stone resulted in a fissure after exposure to the high vacuum, serving as a verification of its presence after 331 days. Once monument 1 achieved 365 days of field testing, another extrusion was removed from the R in "AIR" in a similar process. The SEM images of this sample show similar confirmations as the sample taken at 331 days supporting its durability up to and exceeding one year (Figure 3).

Figure 3. SEM images at 100× (**left**) and 500× (**right**) magnifications of a granite (quartz and feldspar) head stone 365 days after application of formula 11 outdoors and being left to weather. Yellow lines indicate the region of focus in the magnified image on the right.

Images of the raw glass surface show spotting from oils, while the surface coated with formula 11 has minimal distinctions. Glass microscope slides have no surface depressions that air could be trapped in, eliminating the presence of fissures seen in the stone micrographs. A region of the surface which had encased debris and formed a defect was chosen as the remainder of the area was featureless, and both sides are smooth by nature (Figure S15). The raw glass appears dirtier despite both sides being cleaned with dry air prior to imaging.

An example stainless steel sample (histology blade) was inspected by SEM after being dip-coated with formula 12, which includes an ultra-fine edge less than one micron thick (Figure 4 and Figure S16). A slight bevel is visible on the coated surface where the system closely encases the hone, but the overall edge remains well defined. Small indentations are evident in the final magnification; however, there is no evidence of any system defects, and the entirety of the blade edge remains smoother than the uncoated example. This suggests that this system may be used to cover acute angles on the microscopic scale.

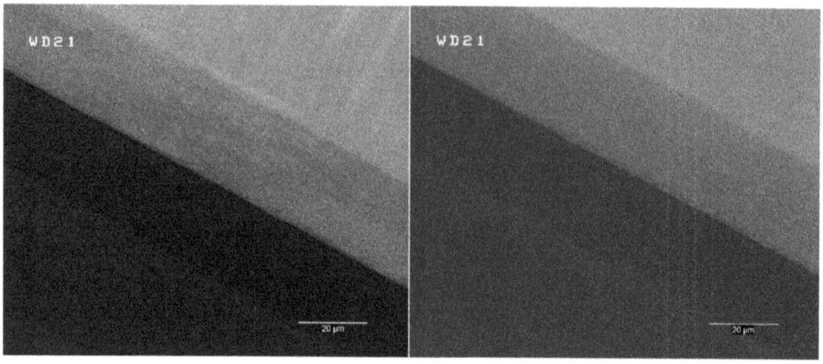

Figure 4. Imaging of stainless-steel histology blades without the manufacturer's coating. The hone of the blade is compared at 1000× magnifications. The raw steel (**left**) and formula 12 dip-coated sample (**right**) both exhibit a submicron edge.

Lastly, the surface interaction of a nitrile glove and wipe-on coating was investigated. Detailed imaging shows how the formula 11 coating covers and fills in many depressions on the surface (Figure S17). At 1000× magnification, a depression in the surface illustrates how thin plates are formed through this application method. While the entire surface is not covered, this may offer additional protection against chemicals that traditionally penetrate nitrile gloves, since the coating exhibited no changes when exposed to some of these reagents detailed below such as THF or DCM.

Examples of coating thickness were analyzed for sprayed and dipped samples. An initial investigation of surface topography and coating thickness was carried out with a Keyence 3D Laser Confocal Microscope and formula 1 applied to stone and steel substrates. Topography mapping of the uncoated and coated sides of a single stone specimen shows a more subtle and consistent surface across the region, which indicates settling of the coating within the depressions of the surface (Figure S18). Coating deposition on a brick surface supports this idea as the solution appears to coat the surface, filling in many imperfections, and getting absorbed into the substrate (Figure S19). Film thickness was assessed using a piece of stainless steel and measuring the distance between the upper and lower scanned surfaces, giving an estimated thickness measurement of 13.19 µm (Figure 5). Next, ten measurements were taken for Formulas 11 and 12 applied to glass slides and used to calculate the average thickness (Table S2). Sprayed samples had an average thickness of 2.76 µm (11) and 10.28 µm (12) while the dip-coated sample had a thickness of 1.35 µm (12), suggesting that this is the ideal application method for a thinner coating. Both measurement methods suggest a similar thickness, however formula 11 is notably thinner even when sprayed, due to the higher ratio of methanol used.

Coating stability was investigated through submersion tests detailed in the methods section. Starting with formula 4, the systems using methanol as a solvent showed no changes in the surface when submerged in water (pH 6.5), synthetic acid rain (pH 4.3), and a sodium hydroxide (pH 12) solution for 24 h each. Prior to formula 4, delamination was observed with water exposure. Chemical resistance was observed in formulas 8–12 when pieces of cured coating were also exposed to small amounts of THF, hexanes, DCM, ethanol, isopropanol, 0.1 M nitric acid, 0.1 M sodium hydroxide, and acetone with no change in sample size, shape, color, or condition. This stability against acetone is an improvement upon many single cure epoxy coatings which it's known to remove. When exposed to a siloxane depolymerization solution containing a fluoride catalyst [42] for four weeks, partial degradation was observed in a sample of formula 10, but the solid was still partially intact, likely due to the thiol-ene and epoxy/amine cross-linking (Figure S20). The overall chemical inertness is most likely due to a combination of the fluoropolymer and the orthogonal reactions which are susceptible to different chemicals and may shield each other when bound together in a network.

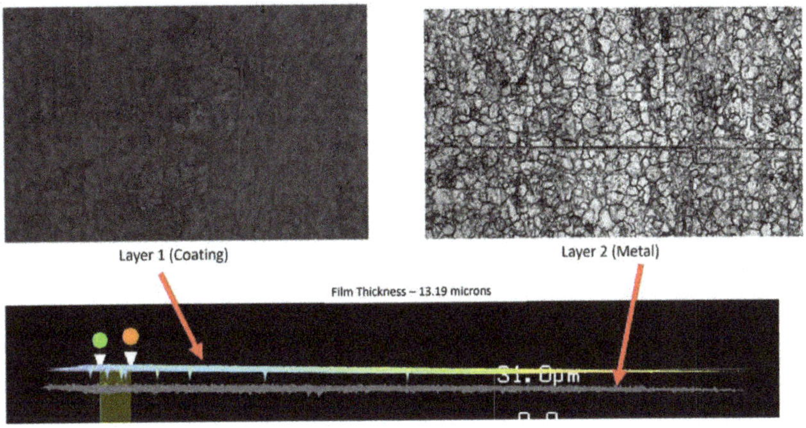

Figure 5. Film thickness analysis performed with a 3D Laser Confocal Microscope which scans the coating surface and sublayer to approximate the difference in distance. Images are of a piece of stainless steel coated with formula 1.

All coatings containing some quantity of fluorocarbon offered graffiti resistant surfaces. Both Sharpie permanent marker and Rust-o-leum outdoor paint can be removed from coated surfaces using a dry Kim-wipe and light force, leaving the coated surface unscathed (Figures S21 and S22, Gif S1, Videos S1 and S2). Ease of removal is correlated with the presence of the perfluoroalkyl silanes which may repel the solvent in the marker and prevent proper adhesion of the paint to the surface. This has a wide range of implications and is highly beneficial if the coating is used on monuments that are subject to vandalism.

Hydrophobicity was built into this system's design mainly to prevent water sitting on surfaces of outdoor structures from soaking into them. This was assessed through the measurement of static contact angles. As mentioned above, the implementation of fluorocarbons increased the overall hydrophobicity observed and was integrated into the system for this reason. The range of contact angles observed in the different variations of all formulas explored was larger than the ideal model system and its lineage (1–13) of formation shown in Table 4; however, all coatings improved upon the contact angles observed on the raw substrates. Coatings on marble had a range between 92.2° and 101.6° with many ideal samples measuring between 96° and 101° throughout the development of the model system. To reduce any effects of surface roughness, some samples were tested on glass microscope slides to ensure consistent comparisons. When applied to glass a maximum contact angle of 105.5° and a minimum of 101.4° were found. The final coatings (11–13) were tested on several different substrates to ascertain their effectiveness on other materials. These results are reasonable when considering formula 12 and 13 which contain ~7% (post cure) of the perfluoroalkyl silane to the pure loading (100% post cure) of the same fluorosilane on marble, glass, and wood which showed static contact angles of 155°, 114°, and 133°, respectively [51]. A 93% reduction in the amount of this compound still yielded 101°, 101°, and 131° on similar substrates. While hydrophobicity is ideal, using high loads of fluorocompounds reduces the overall applicability of the network, especially when there are no adhesive or crosslinking integrations in the system.

Table 4. Average static contact angles of samples on different surfaces over the course of development. Selected samples are part of the progression of the model systems development and were used in the other analysis methods. Standard deviations are given in parenthesis.

Formula No.	Average Contact Angle on Marble	Average Contact Angle on Glass	Average Contact Angle on Brick	Average Contact Angle on Steel	Average Contact Angle on Wood
Uncoated	73.3° (±0.4)	45.8° (±0.5)	Absorbed	58.8° (±5.3)	Absorbed
1	94.9° (±0.8)	101.4° (±0.2)	-	-	-
2	-	105° (±1.2)	-	-	-
3	92.2° (±0.6)	-	-	-	-
4	97° (±1.1)	-	-	-	-
5	99° (±1.1)	-	-	-	-
6	93° (±1.6)	-	-	-	-
7	101.6° (±0.2)	-	-	-	-
8	96° (±1.1)	-	-	-	-
9	100.8° (±0.4)	-	-	-	-
10	96.9° (±0.5)	-	-	-	-
11	100.1° (±0.1)	105.5° (±0.4)	104° (±6.2)	-	-
12	101.5° (±0.1)	101.8° (±0.6)	129.1° (±0.2)	103.7° (±0.4)	131.7° (±0.4)
13	-	104.2° (±0.4)	-	-	-

Brick and steel were investigated as they are used in monument construction and have similar environmental erosion issues like stone. Since bricks are porous, contact angles were taken before, post-cure, and several days after for coating formulas 11 and 12. The non-coated surface completely absorbs water droplets within 30 s (Gif S2). For the first brick formula 11, two days post-cure, it exhibited a contact angle of 103.9° after water sat 10 min on the surface, and at 39 days outdoor, it measured at 94.4° after the coating had worked its way further into the brick (Figure S23, Gif S3). A second brick was coated and inspected four days post-cure, where the contact angle was measured after 10 min had elapsed from deposition and achieved 129.1°, an extremely high angle compared to our other substrates (Figure S23, Gif S4). Stainless steel also showed drastic differences in hydrophobicity after dip-coating with formula 12. The uncoated samples had initial contact angles of 53.5° (1) and 58.8° (S1) on the raw surface, but once coated, measured at 103.5° (2) and 103.7° (S2) in Figures 6 and S24. This indicated that the coating was highly effective at increasing the hydrophobicity of steel surfaces and may add further protection such as oxidative resistance (see below). Nitrile gloves were also investigated by accident to see if additional hydrophobicity could be imparted. Solutions 11 and 12 were wiped onto the surface, and once hardened, the surface was pulled taught. Contact angles were estimated due to the elastic nature of the substrate, where the raw surfaces were ~88.1°, and the coated samples were appraised around 88.8° (formula 11) and 93.4° (formula 12). While the difference in contact angles was not astounding, water droplets ran off the surface easier than the uncoated glove, which indicates this system has potential for use on vinyl surfaces. The effects of adding the coating to wood was investigated as many historical buildings use it as their main construction material. When comparing both raw and formula 12 coated blocks, water begins absorbing into the wood and within 30 s the droplet has a static contact angle under 50° while after 10 min has elapsed the coated sample still repels the water with a contact angle of 131.7° (Video S3). Similar effects were seen with the brick samples, suggesting that porous substrates absorb the solution, enhancing the hydrophobicity of the system.

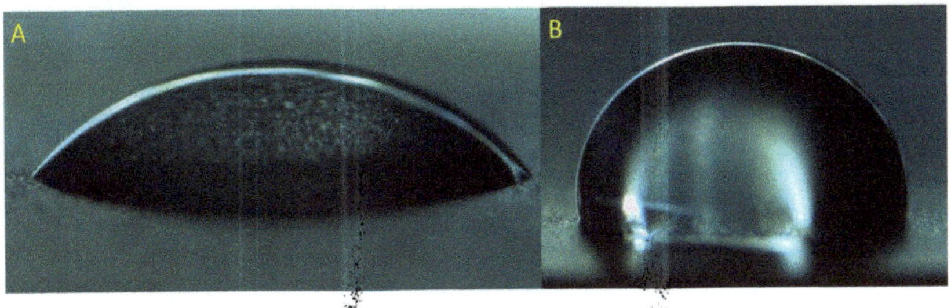

Figure 6. Static water droplets on the surface of raw steel (**A**) and the coated surface with formula 12 (**B**).

Surface assessment was carried out to investigate the hardness and abrasion resistance of the coating. Pencil hardness tests were conducted with samples on various surfaces (marble, glass, steel), achieving a hardness of 9H (~0.5 GPa) on each for every sample assessed after 2 days cured, the highest rating with this method [52]. Both marble and glass microscope slides also exhibited a pencil hardness of 9H, however the graphite was difficult to remove from the stone, still visible on the glass, but easily removed from the coated samples. The final coatings surface is relatively hard compared to other coatings systems such as Axalta's epoxy coating (6H) [53] and an alkoxysilane coating utilizing the same epoxy component as included here (8H) [54]. Abrasion resistance was quantified using a combination of sandpaper and sandblasting techniques and measuring the mass lost from each marble sample. Initial tests with small, coated stone (marble) pieces (average of 3.5 cm × 4 cm) had a maximum mass loss of 3 mg after both smooth (400 grit) and then rough (220 grit) sandpapers (with 1.43 kPa applied force). After exposure to both sandpaper trials, two larger samples (average 8.5 cm × 10 cm) exhibited a total mass loss of 4 mg and 7 mg after 100 rub cycles each. Overall, with a direct abrasive force, the change in surfaces may be considered minimal (207.768 g to 207.764 g for formula 8 and 197.445 g to 197.438 g for formula 9). Marring on these stone samples was observable with a microscope but not the naked eye; however, it is evident when repeated on a transparent coated glass slide. Next, the same samples were sandblasted for thirty seconds at 45 psi, and lost 15 mg and 5 mg, respectively, and damage was visibly evident (Figure 7). This was reinforced by a loss in surface hydrophobicity in the second sample, formula 9, which dropped from 101.5° to 80.3° after sandblasting, suggesting a significant loss of coating (Figure S25). This suggests that while the coating system is rigid and provides some resistance to minor abrasion, it may not be ideal for protecting surfaces in dry, sandy environments that undergo frequent sandstorms without recoating.

Stainless steel samples were put through oxidative stress using a rusting solution (water, vinegar, and sodium chloride). After dip coating, samples were sprayed or submerged (see methods), and both raw and coated surfaces were observed. When sprayed, the solution quickly rusted the raw surface while the coated surface was adequately protected; this is seen in Figure 8, where the top half of the sample was left raw and the lower was coated with formula 12. This suggests that the coating system adequately prevents oxidation when exposed periodically through salt spray (Figure S26). A second rusting test was conducted by submerging a stainless-steel post which was dip-coated directly into the rusting solution (Figure S27). The second image shows the formation of bubbles on the coating surface, but not the raw material after one day. Once 14 days had passed, significant oxidation had occurred on the uncoated surface, but the coated section only had minor rusting and discoloration, indicating the process had been slowed. Monument preservation includes the reduction of oxidation on a variety of surfaces, the coating system demonstrates promise in this regard, especially when the substrate will remain terrestrial.

Figure 7. Images of a marble sample taken with a Zeiss Stemi 2000-C light microscope at 0.65× (**A**) and 2.0× magnification (**B**) before (1) and after sandblasting (2).

Water staining or build up can also be a detriment to coated surfaces, leading to discoloration and calcification over time. Resistance to ion adhesion was tested by partially coating glass slides with formula 12 and using a concentrated synthetic hard water solution with sodium, calcium, and magnesium. Rapid heating resulted in the formation of hard residue on the coated surface, and a dry Kimwipe was used to attempt the removal of it (Gif S5). Using the same force, the calcification easily wipes off the coated surface, but adheres while crumbling apart on the raw surface. The resulting build-up was soft when the same experiment was conducted but dried through slow heating. A dry Kimwipe was used to remove the spots on the uncoated side (Gif S6) and coated side (Gif S7). While the build-up was quickly removed, there was smudging on both sides, which was readily wiped off the coating and remained on the glass. These results suggest that this coating system offers hard water resistance and helps prevent hard water build-up.

Figure 8. Stainless-steel sample before and after seven days of daily exposure to an oxidative rusting solution. The bottom half of the sample was dip coated.

Environmental stability was the final physical characterization investigated. Initial testing was performed indoors to test general thermal stability when exposed to varying circumstances. For both methods, heating and cooling, exposure to high (45 °C) and low (−19 °C) temperatures had no noticeable effect on surface appearance, survivability in submersion tests, and a negligible effect on post-exposure contact angles (± 3°). As previously mentioned, submersion tests were also conducted using a synthetic acid rain solution with a pH of 4.3 to mimic outdoor weather phenomena, and there was no effect on the system.

Long-term stability tests were conducted on a flat rooftop, providing a secure outdoor environment (5 story building in Bowling Green, OH) with limited access to reduce the artificial influence. Daily weather and atmospheric conditions have been recorded since the start of these tests, and samples were inspected periodically to ensure continued success and note any unusual events which may affect testing (see weather log in Supplementary Materials). To gauge the optimal conditions for coating applications, the UV Index was recorded and monitored. This approximates the UV exposure a particular area is forecasted to receive [($mW/m^2/s$)/25 mW]. Successful samples were applied during the summer, and early fall with a UV index exceeding 7.2 (~180 $mW/m^2/s$), compared to the higher power 200 W UVA Lamp used indoors which measures 18.7 mW/cm^2 of power output per second at the used distance. When coated during peak exposure times at a UV Index exceeding 7.2 and a temperature over 21 °C, the sample surface is non-impressionable within 30 min of application as intended. Samples coated at lower UV index ratings and then exposed to higher amounts within two hours saw issues with splitting. This was believed to occur from post hardening thiol-ene reactivity causing stress in the network, which has brought about the suggestion of application with a minimum index of 7 (125 $mW/m^2/s$). The longest-lived sample/marble slab has been in place for 371 days and used a diluted variation of formula 8 (Figure 9). Until around 250 days, it had shown no change to the surface, but as seen in Figure 9D, it has begun to lose the glossy appearance it had; however, the coating is still partially intact and noticeable by touch due to its leathery texture but has lost some hydrophobicity with water droplets spreading on contact.

For upscaling, a granite headstone was cleaned and sanded to remove any existing protective surfaces and then coated using a pressurized pump sprayer which yielded a slightly thicker but even finish (Gif S8). This monument has been in field testing for 371 days and has experienced various environmental conditions including sun, rain, wind, and snow with no notable surface changes. This may be attributed to the hydrophobicity of the surface and good adhesion of the coating, with smaller-scale samples applied in the same manner exhibiting a contact angle of 100.1° on stone and 105.5° on glass (Table 4—Formula 11). Before application, the headstone had a dull gray appearance after being sanded and washed (A), and after application, the surface of the stone became more vivid, enhancing the natural colors of the stone (B), as seen in Figure S28. This change in appearance, accompanied by an increase in the reflectivity of the sun and leathery texture, has continued to indicate the presence and durability of this coating (Figures S28 and S29) in addition to the verification by SEM imaging after 331 days (Figure S14) and at 365 days (Figure 2). A closer look at the edge and etching details emphasizes the difference between the uncoated and protected surfaces as well as the sharpness of detail in the coated surface (Figure 10). Comparing images of the monument before coating and after 365 days had elapsed (Figures S28A and S29) using the CIE76 [50] color difference equation supports the theory that the enhanced detail is from the original stone, and not a defacement. The findings suggest that there is no visible difference between the surface colors on the macroscopic scale with a $\Delta E^*ab = 1$, which indicates alterations are not observable to the naked eye, a requirement for preservation coatings (Figure S30, Formula S1). Two more monuments were coated using formula 12, however these were excluded from stability testing since the first was damaged during nearby construction and the other had partially delaminated due to poor application conditions (wind) causing a thicker coat. Both samples have remained in place for 357 days to gain other data from long-term weathering on this system (ATR-FTIR and TGA). A similar initial enhancement of the surface color is also seen in the two brick samples previously mentioned, but over time this wetted appearance tends to fade (Figures S31 and S32).

Figure 9. Sample before (**A**), after one day (**B**), 24 days (**C**), and 208 days (**D**) after application of formula 8.

3.6. Flame Resistibility

Lastly, flame resistibility of the coating system was tested using two pieces of wood (pine), one raw and one dip coated with formula 12. These samples were burned with a Bunsen burner (>1000 °C) for 3 min and 50 s. Upon turning off the flame source, the coated sample self-extinguished in 11 s while the raw wood continued burning for 1 min and 50 s (Video S4). Inspection of the wood remains show that much of the original shape was retained in the coated sample even with charring evident, while the raw sample became rounded and lost much of its structure (Figure S33). This suggests the coating system has applicability for wooden structures in places prone to fires.

Figure 10. Close up of granite monument edge before coating (**A**), after one day (**B**), 111 days (**C**), 194 days (**D**), and 365 days (**E**) of environmental exposure after application of formula 11.

3.7. Optimal Formulas and Coating Methods

When considering the various tests described above, formulas 11 and 12 were used the most due to 11′s success when applied on the first monument (371 days intact), and 12′s lowered cost with similar attributes. Initial samples of these formula revealed promising contact angles, hardness, thicknesses, abrasion resistances, and graffiti resistances coupled with the stability and adhesion achieved through earlier solutions 1–10. Since these were made in large batches, they were readily used in the pump sprayer and for dip coating larger items. These systems exhibited success with the environmental stability testing due to the addition of a heavier carrier solvent when compared to previous variants, which gave the system more time to react and rearrange before settling on the surface. This discovery is what led to the implementation of solution 13 to lengthen the working time once applied to a surface. This coating was useful as a wipe on coating on a bass boat as shown in

Figure S34, in which after 6 months it is still holding strong. These three formulations show the best overall performance and will be the focus of any future developments and testing.

Ideal application parameters have been established as follows. When applied to surfaces outdoors, the system should be exposed to sunlight when the UV Index exceeds 7 and the temperature is higher than 21 °C. To ensure a complete cure, usage should be limited to days with no precipitation forecast. The indoor application should utilize a UVA Lamp with ~15–20 mW/cm^2 irradiation for 2 min after application. When dip-coated, the lowest edge of the item may need to have excess solution removed. The solution should be stored in an airtight amber bottle in a cool dark location. By following these suggestions, the system will form a non-impressionable surface within thirty minutes of initial UV exposure and stores as a one-pot solution for over six months with repeated usability. Further tool/glass coating examples are shown in Figures S35–S38 to show its utility.

4. Conclusions

Integration of three orthogonal polymeric chemistries has led to the formation of a one-pot coating system that combines each species' (thiol-ene, amine-epoxy, and alkoxysilane) preferred characteristics. By using complementary systems, many of the negative characteristics of the individual components are overcome. Both a radical photo initiator for the rapid thiol-ene reaction and a photo acid generator to accelerate the epoxy/amine and alkoxysilane polymerization greatly reduces the premature reactivity of the components, and, with alcoholic solvents, allows for it to be stored for long periods of time as a single-pot system. A small amount of fluorocarbon also offers just enough hydrophobicity to improve graffiti and chemical resistance. While the ideal formulas (11 and 12) have been stored and reused repeatedly (>once a month) since their creation (371 days and 286 days), the earliest formulas created, including formula 1, are still liquid at 629 days despite having a low solvent load. The ability to store these coating systems as a single unit and initiate multiple orthogonal cures on demand in this manner is unique to itself. This is in contrast to most market-available coatings utilizing one or two curing methods which often have a short shelf life, require separate storage, or have limited reusability. This design allows for the systems to be made in large batches that are readily used in pump sprayers and dip coating applications prior to being resealed and stored for later usage.

Final cured coatings (11 and 12) revealed decent contact angles (>100° for all substrates), excellent hardness (9H), tunable thicknesses (1–12 μm), chemical and graffiti resistances, high thermal stability (~250 °C), and photo stability; all coupled with the enhanced adhesion of the epoxy-amine components. Protected stone monument surfaces exhibit long-term stability and maintain low wettability for >1 year in an unprotected outdoor environment exposed to sun, rain, snow/ice, wind, and temperatures from −19 to 38 °C making it an ideal coating for preserving historical structures. Protection of other substrates are also verified including glass, steel, brick, wood, and plastics, all of which show good coatability, including offering building material stability through flame resistance in wood and environmental resistance in bricks. The combined characteristics detailed above form a series of novel coating systems for application on both porous and non-porous monuments and building materials for protection from a variety of degradation sources.

Future work will investigate substrates and their coated counterparts for evidence of salt decay, crystallization, and how the system effects these processes. Additional studies would look at a variety of new substrates for potential industrial applications, and the surface properties of these systems. Limited success was garnered with recoating, removal, and recyclability efforts which have required the use of different application methods and a combination of organic solvents. Focus will be given to the recyclability of this system back to macromonomers. This will also aid in expediting the removal process of this system and give a better understanding of the underlying chemistries.

5. Patents

A US patent application (63/356,197) has been submitted on this topic before publication.

Supplementary Materials: The following supporting information can be downloaded at: https://www.mdpi.com/article/10.3390/coatings12081098/s1, Figures S1–S38 show additional characterization data including, FTIR, TGA, Contact Angle, SEM, and Surface Profiles, as well as additional application images; Table S1 shows elemental analysis obtained through laser-induced breakdown spectroscopy, Table S2 shows coating thickness data obtained by profilometry; Formula S1 shows the CIE76 formula for color difference analysis; Gifs S1–S8 S: Coating examples on brick, stone, and graffiti removal; Videos S1–S4: Graffiti removal and flame test examples; Daily weather logs are available as "Weather Log 7-20-22.xlsx."

Author Contributions: Conceptualization, J.C.F., C.U.L. and C.B.S.; methodology, J.C.F., C.U.L. and C.B.S.; validation, J.C.F., C.U.L. and C.B.S.; formal analysis, C.B.S.; investigation, C.U.L. and C.B.S.; resources, J.C.F.; data curation, C.U.L. and C.B.S.; writing—draft and editing preparation, J.C.F. and C.B.S.; visualization, C.B.S.; supervision, J.C.F.; project administration, J.C.F.; funding acquisition, J.C.F. All authors have read and agreed to the published version of the manuscript.

Funding: This research was funded by the United States Department of Interior, National Park Service, National Center for Preservation Technology and Training, grant number P20AP00319. Additional support for instrumentation was provided by the Center for Photochemical Sciences and a Faculty Startup Grant at Bowling Green State University.

Institutional Review Board Statement: Not applicable.

Informed Consent Statement: Not applicable.

Data Availability Statement: Raw data for experiments and characterization is available upon request from the corresponding author.

Acknowledgments: J.F., C.S. and C.L. thank the National Park Service for grant funding and support. J.F., C.S. and C.L. thank Bowling Green State University for usage of Chemistry department equipment. J.F. and C.S. thank Jack Cortelezzi, Keyence sales engineer, for usage of the VK-X1000 3D Laser Confocal Microscope. J.F. and C.S. thank Jordon Marich, Keyence sales specialist, for usage of the VHX-7000 Digital Microscope with EA-300. C.S. thanks Nai-hsuan Hu and Herenia Espitia Armenta for their help with photography and equipment maintenance. C.S. thanks Travis Rindler for editorial assistance. J.F. would like to thank Ben Thompson of Mavinate Inc., Jacksonville, FL for testing the coating on real world boat applications.

Conflicts of Interest: The authors declare no conflict of interest.

References

1. Department of Housing Local Government and Heritage Newgrange. Available online: https://www.worldheritageireland.ie/bru-na-boinne/built-heritage/newgrange/ (accessed on 16 February 2022).
2. Tikkanen, A. Britannica, Great Sphinx of Giza. Available online: https://www.britannica.com/topic/Great-Sphinx (accessed on 16 February 2022).
3. Milano Monuments Your Memorial Cost Guide: How to Understand the Costs and Recognize Value. Available online: https://www.milanomonuments.com/your-memorial-cost-guide (accessed on 16 February 2022).
4. Rome Monument High End Monuments. Available online: https://www.romemonuments.com/high-quality-monuments-and-mausoleums (accessed on 16 February 2022).
5. The Viethnam Veterans Memorial Fund About The Wall. Available online: https://www.vvmf.org/About-The-Wall/ (accessed on 21 February 2022).
6. National Park Service U.S. Department of the Interior Washington Monument Construction Timeline. Available online: https://www.nps.gov/wamo/learn/historyculture/monumentconstruction.htm (accessed on 21 February 2022).
7. Washington Info. How Much Did the Washington Monument Cost to Build. Available online: https://www.washingtoncountyinfo.com/the-main-thing-about-washington/how-much-did-the-washington-monument-cost-to-build.html (accessed on 27 February 2022).
8. Stanislaus County Department of Public Works Development Services Division. Preserving Survey Monuments—FAQ. Available online: https://www.stancounty.com/publicworks/pdf/mon_preservation_faq.pdf (accessed on 27 February 2022).
9. Eric, D.; Price Clifford, A. *Epoxy Resins in Stone Consolidation*; Research in Conservation Technical Report Series; Oxford University Press: Oxford, UK, 2010; ISBN 0892362383.
10. Abdellah, M.Y.; Gelany, A.F.; Mohamed, A.F.; Bakr Khoshaim, A. Protection of Limestone Coated with Different Polymeric Materials. *Am. J. Mech. Eng.* **2017**, *5*, 51–57. [CrossRef]

11. Baedecker, P.A.; Reddy, M.M.; Reimann, K.J.; Sciammarella, C.A. Effects of Acidic Deposition on the Erosion of Carbonate Stone—Experimental Results from the U.S. National Acid Precipitation Assessment Program (NAPAP). *Atmos. Environ. Part B Urban Atmos.* **1992**, *26*, 147–158. [CrossRef]
12. Li, Q.; Zhang, B.; Yang, X.; Ge, Q. Deterioration-Associated Microbiome of Stone Monuments: Structure, Variation, and Assembly. *Appl. Environ. Microbiol.* **2018**, *84*, e02680-17. [CrossRef] [PubMed]
13. Terlikowski, W.; Sobczyńska, E.; Gregoriou-Szczepaniak, M.; Wasilewski, K. Natural and Synthetic Polymers Used in the Preservation of Historical Stone Buildings. *IOP Conf. Ser. Mater. Sci. Eng.* **2019**, *661*, 012135. [CrossRef]
14. Sadat-Shojai, M.; Ershad-Langroudi, A. Crystallization Behavior of Poly(ε-Caprolactone)/Layered Double Hydroxide Nanocomposites. *J. Appl. Polym. Sci.* **2009**, *112*, 2535–2551. [CrossRef]
15. Wright, J.D.; Sommerdijk, N.A.J.M. *Sol-Gel Materials: Chemistry and Applications*, 1st ed.; CRC Press: Boca Raton, FL, USA, 2000; ISBN 9789056993269.
16. Wang, D.; Bierwagen, G.P. Sol-Gel Coatings on Metals for Corrosion Protection. *Prog. Org. Coat.* **2009**, *64*, 327–338. [CrossRef]
17. Park, H.Y.; Kloxin, C.J.; Scott, T.F.; Bowman, C.N. Stress Relaxation by Addition-Fragmentation Chain Transfer in Highly Cross-Linked Thiol-Yne Networks. *Macromolecules* **2010**, *43*, 10188–10190. [CrossRef]
18. Monde, T.; Fukube, H.; Nemoto, F.; Yoko, T.; Konakahara, T. Preparation and Surface Properties of Silica-Gel Coating Containing Branched-Polyfluoroalkylsilane. *J. Non. Cryst. Solids* **1999**, *246*, 54–64. [CrossRef]
19. Daoud, W.A.; Xin, J.H.; Tao, X. Synthesis and Characterization of Hydrophobic Silica Nanocomposites. *Appl. Surf. Sci.* **2006**, *252*, 5368–5371. [CrossRef]
20. Gautam, R.; Ian Geniza, I.; Iacono, S.T.; Friesen, C.M.; Jennings, A.R. Perfluoropyridine: Discovery, Chemistry, and Applications in Polymers and Material Science. *Molecules* **2022**, *27*, 1616. [CrossRef]
21. Stewart, K.A.; Shuster, D.; Leising, M.; Coolidge, I.; Lee, E.; Stevens, C.; Peloquin, A.J.; Kure, D.; Jennings, A.R.; Iacono, S.T. Synthesis, Characterization, and Thermal Properties of Fluoropyridyl-Functionalized Siloxanes of Diverse Polymeric Architectures. *Macromolecules* **2021**, *54*, 4871–4879. [CrossRef]
22. Ershad-Langroudi, A.; Mai, C.; Vigier, G.; Vassoille, R. Hydrophobic Hybrid Inorganic-Organic Thin Film Prepared by Sol-Gel Process for Glass Protection and Strengthening Applications. *J. Appl. Polym. Sci.* **1997**, *65*, 2387–2393. [CrossRef]
23. Xu, F.; Zeng, W.; Li, D. Recent Advance in Alkoxysilane-Based Consolidants for Stone. *Prog. Org. Coat.* **2019**, *127*, 45–54. [CrossRef]
24. Gordon, N. Silicone Automobile Finish Protectant. U.S. Patent 4,997,478, 5 March 1991. pp. 1–4.
25. Mark, J.E. Overview of Siloxane Polymers. *ACS Symp. Ser.* **1999**, *729*, 1–10. [CrossRef]
26. Wheeler, G. *Alkoxysilanes and the Consolidation of Stone*; Getty Publications: Los Angeles, CA, USA, 2005; Volume 46, ISBN 9780892368150/0892368152.
27. Heshmati, M.R.; Amiri, S.; Hosseini-Zori, M. Synthesis and Characterization of Superhydrophobic-Superoleophobic and Anti-Corrosion Coatings via Sol-Gel Process. *Open J. Org. Polym. Mater.* **2020**, *10*, 1–15. [CrossRef]
28. Becherini, F.; Lucchi, E.; Gandini, A.; Barrasa, M.C.; Troi, A.; Roberti, F.; Sachini, M.; Di Tuccio, M.C.; Arrieta, L.G.; Pockelé, L.; et al. Characterization and Thermal Performance Evaluation of Infrared Reflective Coatings Compatible with Historic Buildings. *Build. Environ.* **2018**, *134*, 35–46. [CrossRef]
29. Laine, R.M.; Furgal, J.C.; Doan, P.; Pan, D.; Popova, V.; Zhang, X. Avoiding Carbothermal Reduction: Distillation of Alkoxysilanes from Biogenic, Green, and Sustainable Sources. *Angew. Chemie Int. Ed.* **2016**, *55*, 1065–1069. [CrossRef]
30. Furgal, J.C.; Laine, R.M. Nucleophilic Attack of R-Lithium at Tetrahedral Silicon in Alkoxysilanes. An Alternate Mechanism. *Bull. Chem. Soc. Jpn.* **2016**, *89*, 705–725. [CrossRef]
31. Furgal, J.C.; Lenora, C.U. Green Routes to Silicon-Based Materials and Their Environmental Implications. *Phys. Sci. Rev.* **2020**, *5*. [CrossRef]
32. Galhenage, T.P.; Stafslien, S.J.; Skaja, A.; Webster, D.C. Durable Siloxane-Polyurethane Coatings for Mitigating Freshwater Mussel Fouling. *Biofouling* **2022**, *38*, 260–270. [CrossRef]
33. Benda, J.; Stafslien, S.; Vanderwal, L.; Finlay, J.A.; Clare, A.S.; Webster, D.C. Surface Modifying Amphiphilic Additives and Their Effect on the Fouling-Release Performance of Siloxane-Polyurethane Coatings. *Biofouling* **2021**, *37*, 309–326. [CrossRef] [PubMed]
34. Yu, Z.; Cui, A.; Zhao, P.; Wei, H.; Hu, F. Preparation and Properties Studies of UV-Curable Silicone Modified Epoxy Resin Composite System. *J. Appl. Biomater. Funct. Mater.* **2018**, *16*, 170–176. [CrossRef] [PubMed]
35. Tuteja, A.; Choi, W.; McKinley, G.H.; Cohen, R.E.; Rubner, M.F. Design Parameters for Superhydrophobicity and Superoleophobicity. *MRS Bull.* **2008**, *33*, 752–758. [CrossRef]
36. Tuteja, A.; Choi, W.; Ma, M.; Mabry, J.M.; Mazzella, S.A.; Rutledge, G.C.; McKinley, G.H.; Cohen, R.E. Designing Superoleophobic Surfaces. *Science* **2007**, *318*, 1618–1622. [CrossRef]
37. Ruffolo, S.A.; La Russa, M.F. Nanostructured Coatings for Stone Protection: An Overview. *Front. Mater.* **2019**, *6*, 147. [CrossRef]
38. Crupi, V.; Fazio, B.; Gessini, A.; Kis, Z.; La Russa, M.F.; Majolino, D.; Masciovecchio, C.; Ricca, M.; Rossi, B.; Ruffolo, S.A.; et al. TiO2–SiO2–PDMS Nanocomposite Coating with Self-Cleaning Effect for Stone Material: Finding the Optimal Amount of TiO2. *Constr. Build. Mater.* **2018**, *166*, 464–471. [CrossRef]
39. De Ferri, L.; Lottici, P.P.; Lorenzi, A.; Montenero, A.; Salvioli-Mariani, E. Study of Silica Nanoparticles—Polysiloxane Hydrophobic Treatments for Stone-Based Monument Protection. *J. Cult. Herit.* **2011**, *12*, 356–363. [CrossRef]

40. Hu, N.H.; Lenora, C.U.; May, T.A.; Hershberger, N.C.; Furgal, J.C. In Situ Formed Methyl- Co -(Bis-R) Silsesquioxane Based Polymer Networks with Solvent Controlled Pore Size Distributions and High Surface Areas. *Mater. Chem. Front.* **2020**, *4*, 851–861. [CrossRef]
41. Sulaiman, S.; Bhaskar, A.; Zhang, J.; Guda, R.; Goodson, T.; Laine, R.M. Molecules with Perfect Cubic Symmetry as Nanobuilding Blocks for 3-D Assemblies. Elaboration of Octavinylsilsesquioxane. Unusual Luminescence Shifts May Indicate Extended Conjugation Involving the Silsesquioxane Core. *Chem. Mater.* **2008**, *20*, 5563–5573. [CrossRef]
42. Rupasinghe, B.; Furgal, J.C. Full Circle Recycling of Polysiloxanes via Room-Temperature Fluoride-Catalyzed Depolymerization to Repolymerizable Cyclics. *ACS Appl. Polym. Mater.* **2021**, *3*, 1828–1839. [CrossRef]
43. Wahab, G.M.A.; Gouda, M.; Ibrahim, G. Study of Physical and Mechanical Properties for Some of Eastern Desert Dimension Marble and Granite Utilized in Building Decoration. *Ain Shams Eng. J.* **2019**, *10*, 907–915. [CrossRef]
44. Malaga-Starzec, K.; Åkesson, U.; Lindqvist, J.E.; Schouenborg, B. Microscopic and Macroscopic Characterization of the Porosity of Marble as a Function of Temperature and Impregnation. *Constr. Build. Mater.* **2006**, *20*, 939–947. [CrossRef]
45. Sousa, L.; Barabasch, J.; Stein, K.-J.; Siegesmund, S. Characterization and Quality Assessment of Granitic Building Stone Deposits: A Case Study of Two Different Portuguese Granites. *Eng. Geol.* **2017**, *221*, 29–40. [CrossRef]
46. Schild, M.; Siegesmund, S.; Vollbrecht, A.; Mazurek, M. Characterization of Granite Matrix Porosity and Pore-Space Geometry by in Situ and Laboratory Methods. *Geophys. J. Int.* **2001**, *146*, 111–125. [CrossRef]
47. TWC Product and Technology LCC The Weather Channel. Available online: https://weather.com/ (accessed on 18 June 2021).
48. United States Environmental Protection Agency UV Index Search. Available online: https://www.epa.gov/enviro/uv-index-search (accessed on 18 June 2021).
49. WillyWeather USA UV Forecast. Available online: https://uv.willyweather.com/ (accessed on 18 June 2021).
50. Bala, R.; Sharma, G. *Digital Color Imaging Handbook*, 1st ed.; Sharma, G., Ed.; CRC Press: Boca Raton, FL, USA, 2003; ISBN 0-8493-0900-X.
51. Adamopoulos, F.G.; Vouvoudi, E.C.; Achilias, D.S.; Karapanagiotis, I. Fluorosilane Water-Repellent Coating for the Protection of Marble, Wood and Other Materials. *Heritage* **2021**, *4*, 2668–2675. [CrossRef]
52. Chen, Z.; Wu, L.Y.L. *Chapter 14—Scratch Damage Resistance of Silica-Based Sol–Gel Coatings on Polymeric Substrates*; Friedrich, K., Schlarb, A.K.B.T.-T.P.N., Second, E., Eds.; Butterworth-Heinemann: Oxford, UK, 2013; pp. 467–511, ISBN 978-0-444-59455-6.
53. Axalta Coating Systems. Epoxy Technical Summary. *Alesta Technical Guide*; Houston, TX, USA, 2014. Available online: https://www.axalta.com/content/dam/NA/HQ/Public/Powder%20Coatings/Documents/Marketing/EPOXY%20TECHNICAL%20SUMMARY.pdf (accessed on 6 July 2022).
54. Qiu, Z.; Lin, H.; Zeng, L.; Liang, Y.; Zeng, C.; Hong, R. Ultra-Scratch-Resistant, Hydrophobic and Transparent Organosilicon-Epoxy-Resin Coating with a Double Cross-Link Structure. *Appl. Sci.* **2022**, *12*, 4854. [CrossRef]

coatings

MDPI

Article

A Durable Nano-SiO_2-TiO_2/Dodecyltrimethoxysilane Superhydrophobic Coating for Stone Protection

Meiman Peng [1,2,3], Liqin Wang [1,2,*], Lang Guo [3], Jinyi Guo [3], Liping Zheng [3,4], Fuwei Yang [3], Zhuang Ma [5] and Xing Zhao [3,*]

[1] China-Central Asia "the Belt and Road" Joint Laboratory on Human and Environment Research, Xi'an 710127, China
[2] Key Laboratory of Cultural Heritage Research and Conservation, Northwest University, Ministry of Education, Xi'an 710127, China
[3] School of Cultural Heritage, Northwest University, Xi'an 710127, China
[4] School of History and Society, Chongqing Normal University, Chongqing 401331, China
[5] Leibniz-Institut für Katalyse e.V., Albert-Einstein-Str. 29a, D-18059 Rostock, Germany
* Correspondence: wangliqinnwuedu@163.com (L.W.); zhaoxing@nwu.edu.cn (X.Z.)

Abstract: Water can trigger freeze–thaw cycles, acid rain corrosion, and microbial colonisation, all of which destroy stone. Water is one of the most influential factors in the destruction of outdoor stone heritage. Therefore, materials with excellent hydrophobic properties and durability are urgently required to effectively retard long-term stone weathering. In this study, two nanoparticles, TiO_2 and SiO_2, were used to modify dodecyltrimethoxysilane (DTMS), a waterproof coating commonly used for stone heritage protection, to fabricate nanocomposite superhydrophobic coatings. The micromorphology, water repellence (water contact angle and capillary water absorption), suitability to protect stone heritage (color change and water vapor permeability), and durability (thermal, light, and chemical stability) of DTMS and nanocomposite coatings were evaluated. The scanning electron microscope (SEM) images revealed that adding 0.5% (*w/w*) SiO_2 produced nanoscale roughness on the sandstone surface, leading to superhydrophobicity. The results of ultraviolet -visible (UV–Vis) spectrophotometer showed that adding 0.01% TiO_2 shielded more than 90% of UV light but accelerated the decrease in the contact angle under UVA irradiation. The addition of SiO_2 was able to avoid the detrimental effect of TiO_2 under UV light. The thermogravimetric analysis (TGA) results showed that both SiO_2 and TiO_2 nanoparticles improved the thermal stability of the coatings. In particular, the fabricated nanocomposite coating, SiO_2 and TiO_2 co-modified DTMS, had excellent water repellence, low color change and outstanding durability, and retained about 85% of the water vapor permeability of the stone, showing promise for stone protection.

Keywords: stone protection; SiO_2 nanoparticle; TiO_2 nanoparticle; superhydrophobic coatings; durability; dodecyltrimethoxysilane

Citation: Peng, M.; Wang, L.; Guo, L.; Guo, J.; Zheng, L.; Yang, F.; Ma, Z.; Zhao, X. A Durable Nano-SiO_2-TiO_2/Dodecyltrimethoxysilane Superhydrophobic Coating for Stone Protection. *Coatings* **2022**, *12*, 1397. https://doi.org/10.3390/coatings12101397

Academic Editor: Ioannis Karapanagiotis

Received: 30 August 2022
Accepted: 23 September 2022
Published: 25 September 2022

1. Introduction

Sandstone, which has a porous structure, is an important component of stone heritage worldwide [1–5]. Outdoor stone heritage is weathered by temperature, humidity, light, polluting gases, etc., with water being the dominant factor influencing weathering. Stones are damaged by dry–wet cycles, freeze–thaw cycles, salt, and biodeterioration because of the presence of water. Controlling this single factor, water, may effectively inhibit damage to stone heritage and its associated components [6].

Fluoropolymers [7,8], siloxane coatings [9], and acrylic polymers [10] are the surface materials most studied and applied to protect stone. Due to their low surface energy, these organic coatings are hydrophobic. Hydrophobic materials, having a static contact angle larger than 150° and low contact angle hysteresis (usually a roll-off angle of less than 10°), are generally referred to as superhydrophobic materials. The water-repellence ability of superhydrophobic materials enables them to effectively reduce the effect of water erosion, and the low roll-off angle of the surfaces effectively decreases the deposition of contaminants and microorganisms [11], showing potential for the protection of stone artefacts. Obtaining superhydrophobicity requires substances with appropriate roughness and low surface energy [12–14]. A nanocomposite coating, usually prepared with organic coatings and nanoparticles, is currently one of the most used superhydrophobic materials, where the organic component has low surface energy and the nanoparticles provide a rough structure.

Silicones are widely used, low-surface-energy substances that are chemically stabile because of the high-strength Si–O bond. For example, dodecyltrimethoxysilane (DTMS) has been used for alloy anticorrosion [15,16], fabric waterproofing [17], and the protection of historic buildings [18,19] and sandstone [20]. In China, DTMS has been used as a water-repellent coating for the stone artefacts of the Qianling tomb, Maoling tomb, Banpo site, Big Wild Goose Pagoda in Shaanxi Province [21], and Goguryeo site in Jilin Province [22]. Silicones, specifically DTMS, can feasibly be used for conserving stone artefacts. However, its light resistance is poor. After 800 h of UVB irradiation, the water contact angle of a stone surface dropped to 0° [23]. Furthermore, its waterproofing and durability need to be enhanced.

Nanomaterials had initial applications in heritage conservation [24,25]. Organic coatings and nanoparticles are combined to prepare nanocomposite organic coatings with excellent properties. The introduction of nanomaterials may improve the properties of organic coatings, such as their water absorption, thermo–mechanical properties, substrate adhesion, UV and chemical stability, wear resistance [26], and self-cleaning [27], which is expected to improve the durability and water repellence of DTMS, thus further expanding the application of silicones in the field of cultural heritage.

Some researchers have studied the use of single-nanoparticle-modified coatings for the protection of stone artefacts. SiO_2 and TiO_2 nanoparticles are two of the nanoparticles most used for nanocomposite organic coatings to protect cultural heritage [28]. Facia [29] added SiO_2 nanoparticles to siloxane, simply producing superhydrophobic nanocomposite coatings in situ on a sandstone building substrate. TiO_2 nanoparticles [30] were also used to fabricate superhydrophobic nanocomposite coatings, giving photo-catalytic activity and self-cleaning properties [31]. Nano-Al_2O_3 and SnO_2 [30] were added to siloxane to produce superhydrophobic films to protect outdoor cultural heritage assets. Chatzigrigoriou [32] used dispersions of $Ca(OH)_2$ nanoparticles in siloxane emulsions to produce superhydrophobic coatings for marble protection, and the color change was 3.76, which was accepted for conservation purposes. In particular, ZnO nanoparticles [33] and Ag nanoparticles [34] were added to silicon-based consolidant/water-repellent materials, exerting biocide activity. It is worth mentioning that Karapanagiotis had done plenty of research about superhydrophobic materials for the conservation of natural stone and recently presented a detailed review [35]. These works have significantly developed the application of nanocomposite coatings in the field of stone heritage conservation.

However, less research has been done on the durability of nanocomposite coatings, and to the best of our knowledge, cases of the use of multiple-nanoparticle-modified organic coatings for cultural heritage protection are uncommon. As such, in this study, we modified DTMS with nano-SiO_2 and TiO_2 to fabricate superhydrophobic materials for application in sandstone protection. Their hydrophobicity, applicability, and durability for protecting sandstone-based cultural heritage were comprehensively studied. We investigated the effects of adding two kinds of nanoparticles, SiO_2 and TiO_2, separately and simultaneously, to thoroughly understand the role of nano-SiO_2 and TiO_2. Our findings contribute to the understanding of nanocomposite organic coatings and broaden the application of nanocomposites in the field of cultural heritage protection, providing more diverse materials for the protection of outdoor stone heritage.

2. Experimental

2.1. Preparation of Stone Samples

Red sandstone was acquired from the Daming Place Building Material Market, Xi'an. It was mainly composed of quartz, calcite, and feldspars, which were confirmed by XRD (see Supplementary Materials Figure S1 for details). The sandstone surface was cleaned with distilled water and then dried at 110 °C prior to the experiments. The specifications of the samples varied from test to test. Figure 1 shows the sample blocks used for different tests.

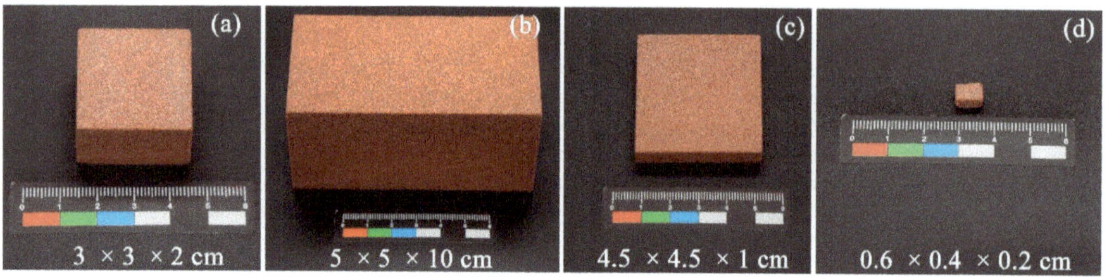

Figure 1. Stone samples used for the tests: (**a**) contact angle, color, thermal ageing, UVA ageing, chemical stability; (**b**) water absorption by capillarity; (**c**) water vapor permeability; (**d**) SEM.

2.2. Preparation and Application of Coatings

Dodecyltrimethoxysilane (DTMS, Sinopharm, Beijing, China), OP-10 (Emulsifier, Aladdin, Shanghai, China), and iso-propanol (Tianjin Fuyu, China) were used as received. The silicon dioxide nanoparticles (fumed powder, Aladdin, Shanghai, China) had an average particle size of 7~40 nm and a specific surface area of 150 m^2/g. The titanium dioxide nanoparticles (P25, Macklin, Shanghai, China) had an average particle size of 20 nm. The P25 TiO_2 consisted of two crystalline forms, anatase and rutile, in a 4:1 ratio. The nanoparticles were dried at 110 °C for 12 h before use.

Table 1 shows the formulations of the four prepared materials. The solutions were dispersed by using a high-speed shear dispersion tester for 5 min (6000 rpm) and then ultrasonicated for 30 min (36 kHz, 12 °C). The nanocomposites show excellent properties only if the nanoparticles are well-dispersed [36].

Table 1. Coating formulations in percent (*w/w*). D, dodecyltrimethoxysilane (DTMS); DT and DS, TiO_2- and SiO_2-modified DTMS, respectively; DST, TiO_2 and SiO_2 comodified DTMS, where iso-propanol was the solvent, OP-10 was the emulsifier, and the concentrations of TiO_2 and SiO_2 were 0.01% (*w/w*) and 0.5% (*w/w*), respectively.

Product	DTMS	Nano-TiO_2	Nano-SiO_2	Iso-Propanol	OP-10
D	10	-	-	90	-
DT	10	0.01	-	90	0.2
DS	10	-	0.5	90	0.2
DST	10	0.01	0.5	90	0.2

The solutions were applied dropwise to the stone surfaces using a Dlab graduated pipette (accuracy ± 1 μL, Beijing, China). One side of the sample was treated twice with the nanocomposite dispersion solutions, with 40 $\mu L/cm^2$ each time, in an interval of 1 h and a cumulative dosage of 80 $\mu L/cm^2$. The treated stones were subjected to 80 °C for two days and then to room temperature for more than two days.

2.3. Characterisation Tests

2.3.1. Morphological Observation

The sample surface micromorphology was determined using a field-emission scanning electron microscope (FESEM; Thermo scientific, Apros S, Waltham, MA, USA) with a 10 kV high voltage and an 11 mm work distance.

2.3.2. Waterproofing Test

Water Contact Angle Test

A static contact angle measuring instrument (Chenghui, JGW-360, Chengde, China) was used for the test. To ensure repeatability, the sample was measured 10 s after placing a 5 μL drop of water on the treated sample surface. Each sample was measured seven times, and we reported the average value. The roll-off angle was measured with a roll-off angle platform.

Water Absorption Capillarity Test

Flat cotton (total thickness of approximately 7 mm) was placed in a clean plate, and distilled water was added until the cotton was completely soaked, ensuring the water level did not exceed the upper surface of the cotton and remained constant during the test [37]. Each sample was weighed and was placed on top of the cotton to absorb water until the weight was constant. The amount of water absorbed per unit area Q_i ($kg{\cdot}m^{-2}$) at time t_i (s) was calculated as follows:

$$Q_i = \frac{m_i - m_0}{A} \quad (1)$$

where m_0 is the mass of the specimen at time t_0, in kilograms; m_i is the mass of the specimen at time t_i, in kilograms; A is the area of the specimen in contact with the bedding layer (cotton), in meters squared.

After 72 h of treatment, the weight was basically constant. The water absorption inhibition efficiency (WIE) was calculated as follows [38]:

$$\text{WIE } \% = \frac{m_{un} - m_t}{m_{un}} \times 100 \quad (2)$$

where m_{un} and m_t are the amounts of water absorbed after 72 h of treatment by the untreated sample (the red sandstone) and by the sample treated with waterproofing coatings, respectively.

2.3.3. Suitability Analysis

Colorimetric Measurements

The appearance of the sample was characterized using a colorimeter (X-rite, VS 450, MI, USA). We adopted the CIE $L^*a^*b^*$ color system, where L^* indicates luminosity, 0 indicates black, and 100 indicates white; a^*, which ranges from positive to negative, indicates colors from red to green; b^*, which ranges from positive to negative indicates colors from yellow to blue. The color difference ΔE was calculated as follows:

$$\Delta E = \sqrt{(\Delta L^*)^2 + (\Delta a^*)^2 + (\Delta b^*)^2} \quad (3)$$

where ΔL^*, Δa^*, and Δb^* are the variations in the L^*, a^*, and b^* values of the samples before and after treatment, respectively. Each sample was measured four times, and we reported the average.

Water Vapor Permeability Test

The stones that were both untreated and treated with coatings were subjected to a water vapor permeability test, following UNI EN 15803 [39]. Each cup was filled with equal amounts of K_2SO_4-saturated solution to regulate humidity at 97%. The devices were sealed with paraffin wax and were then placed in a constant-temperature and -humidity chamber at 25 °C and 50% humidity. The devices were weighed every 48 h. The variation in weight, Δm, was calculated as the difference between the weight of the apparatus (in kg) at times t_0 and t_i. The water vapor flow rate through the specimen in the unit of time, G kg·s^{-1}), was calculated as the slope of the Δm vs. t curve. The water vapor permeability, δp (kg·m^{-1}·s^{-1}·Pa^{-1}), was calculated as follows:

$$\delta p = \frac{G}{A \times \Delta_{P_v}} \times D \quad (4)$$

where A is the test surface area, in square meters; Δ_{P_v} is the water vapor pressure difference across the samples, in Pascals; D is the thickness of the samples, in meters.

2.3.4. Durability Analysis

Stone relics, especially outdoors, are inevitably weathered by heat, ultraviolet rays, and some acidic and alkaline chemicals. Therefore, we tested the durability of the coating from three aspects:

Thermal Stability Test

The samples were placed in an oven at 180 °C to accelerate thermal ageing. The contact angles were periodically measured until the contact angle was below 60°, for a total ageing time of 252 h.

To study the difference in the thermal stability between DTMS before and after modifications, 5 mL of the nanocomposite solution was mixed with 1 g of sandstone powder. The mixture (6~7 mg) after completely curing was obtained for thermogravimetric analysis using a simultaneous thermal analyzer (Mettler Toledo, TGA-DSC3+, Zurich, Switzerland) in a nitrogen atmosphere with a flow rate of 20 mL·min^{-1} and a heating rate of 10 °C·min^{-1}.

UV Shielding Test

Of the UV radiation produced by the sun and received by the Earth, almost all the short-wave UVC (200–280 nm) is absorbed by the ozone layer, and about 2%–5% of the medium-wave UVB (280–320 nm) reaches the Earth. The rest of the UV radiation is long-wave UVA (320–400 nm). To simulate the UV radiation damage of outdoor stone artefacts, UVA was chosen as the light source for testing.

The samples were placed in a UV ageing chamber for ageing, where 5 identical UVA lamps (Philips, 8 W 365 nm) were installed. The sample surface was 16 cm from the base of the lamps with a UV irradiance (365 nm) of 10.35 W/m^2. The contact angles were measured every 24 h until the contact angle was below 60°, for a total ageing time of 144 h. The sample position was periodically changed during the experiment to ensure uniform ageing.

The ability of the added nanoparticles to shield against UV light was studied by using a UV–Visible spectrophotometer (Hitachi, U-2001, Tokyo, Japan) with a scanning speed of 200 nm/min. Approximately 4 mL of the solution prepared as described in Section 2.2 was tested. Iso-propanol as the reference solution and a 1 cm thick quartz cuvette were used.

Chemical Durability Test

The samples were immersed in pH 1, 7 (ultrapure water), or 13 solutions for 24 h. The contact angles of the samples before and after treatment were measured.

3. Results and Discussion

3.1. Water-Repellence Performance

3.1.1. Contact Angle

The static contact angles of the samples are reported in Table 2, where S represents untreated sandstone. These images are shown in Figure 2. DTMS (Figure 2a) showed suitable water resistance with a contact angle of 126.5°, and the addition of 0.01% TiO_2 to DTMS (Figure 2b) produced an insignificant improvement in the contact angle of DTMS. Adding 0.5% SiO_2 to DTMS (Figure 2c,d) resulted in a substantial increase in the contact angle beyond 150°, with a roll-off angle of less than 10° measured using a roll-off angle platform. Because the static contact angle was larger than 150° and the roll-off angle was less than 10°, the fabricated nanocomposite DS and DST coatings were superhydrophobic materials.

Table 2. Average values of the static contact angles (θ) and water vapor permeability (δp).

Sample	θ (°)	δp (10^{-11} kg·(m·s·Pa)$^{-1}$)	RP (%)
S	0	1.34 ± 0.06	100.00
D	126.5 ± 6.4	1.13 ± 0.04	84.17
DT	129.3 ± 3.4	1.10 ± 0.05	82.15
DS	152.5 ± 3.5	1.12 ± 0.07	83.68
DST	152.1 ± 2.6	1.14 ± 0.04	84.83

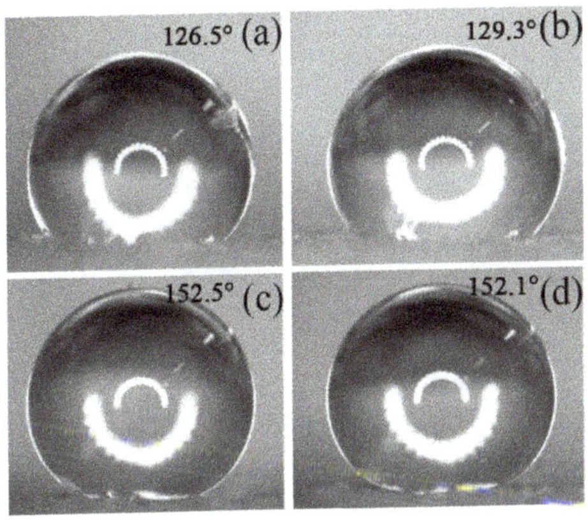

Figure 2. Static contact angle images: (**a**) D; (**b**) DT; (**c**) DS; (**d**) DST.

3.1.2. Water Absorption by Capillarity

The amount of water absorbed by the samples per unit of area Q_i over time is reported in Figure 3. Compared with the untreated samples, the samples treated with DTMS and nanocomposite coatings all had a WIE higher than 92% after 72 h of absorption, showing excellent water-repellence performance.

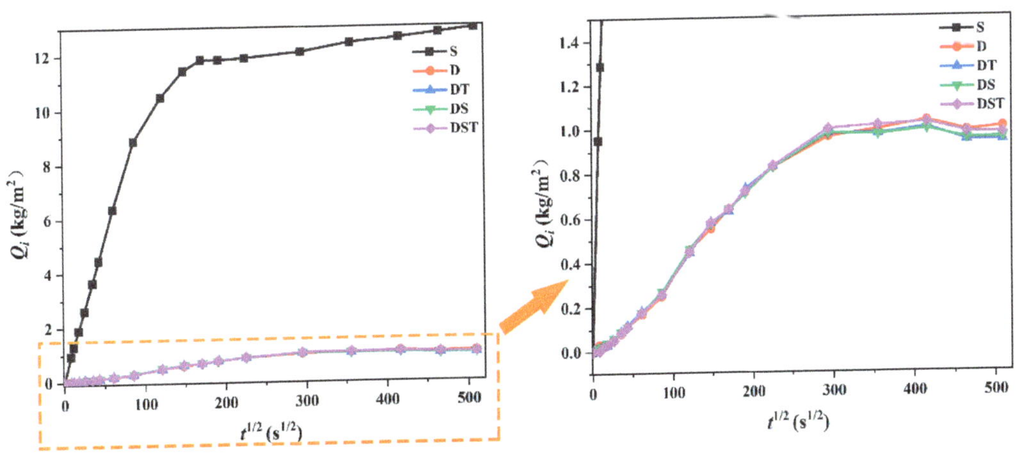

Figure 3. Capillary water-absorption curves.

3.2. Micromorphology

The SEM images of the samples (Figure 4) showed that the original sandstone surface (Figure 4a) was relatively rough. After coating with DTMS, a relatively smooth and low-surface-energy surface formed (Figure 4b), and the latter mainly imparted hydrophobicity to the surface. After the addition of TiO_2 nanoparticles to DTMS, a low content (0.01% *w/w*) of TiO_2 nanoparticles was sporadically distributed on the stone surface (Figure 4c), which was unable to produced sufficient roughness, thereby minimally affecting the contact angle of DTMS. The SiO_2 nanoparticles (0.5% *w/w*) were able to more uniformly gather on the stone surface, forming a rough structure (Figure 4d,e). When water drops fell on the DS and DST surfaces, the air was trapped in the void between the water and rough structure formed by the accumulation of nanoparticles, and the drops thus easily rolled off, which is in accordance with the Cassie–Baxter model [13,14].

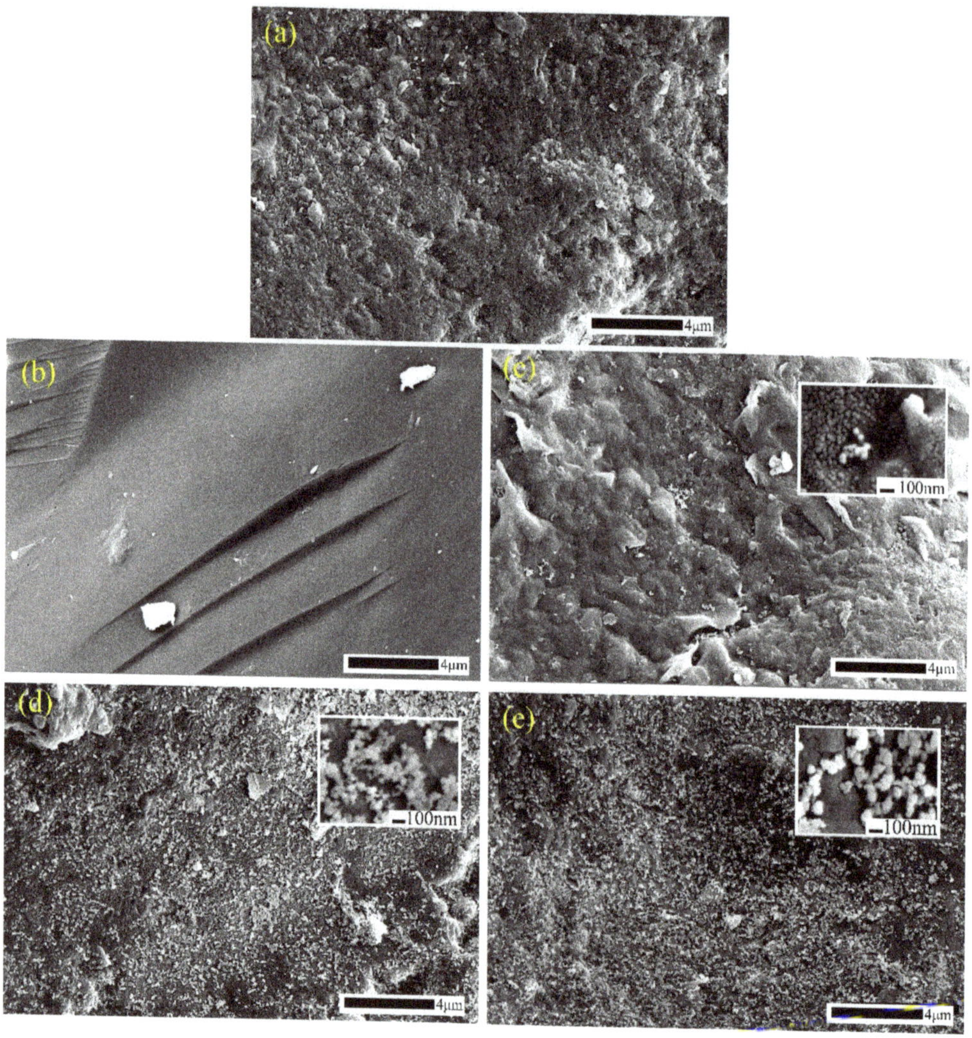

Figure 4. SEM images: (**a**) S, (**b**) D, (**c**) DT, (**d**) DS, and (**e**) DST. A higher-magnification image is depicted in (**c**,**d**), and (**e**) in the upper right corner.

It has been reported that different types of nanoparticles can form rough structures to achieve superhydrophobicity [30]. For elevated particle concentration, the wettabilities of the surfaces were not affected by the original particle sizes of the nanomaterials [40,41]. Nevertheless, adequate roughness of the substrate surface at the high nanoparticle concentration is a key condition for achieving superhydrophobicity [40]. Therefore, adding low concentrations of TiO_2 nanoparticles had little effect on the contact angle of DTMS, while high concentrations of SiO_2 nanoparticles greatly increased the contact angle.

According to the SEM image in Figure 4e, the DST particle size (Figure 5) was calculated as almost 70% in the range of 60~90 nm, with an average particle size of 81 nm, indicating that the nanoparticles were well-dispersed and formed a rough nanoscale structure.

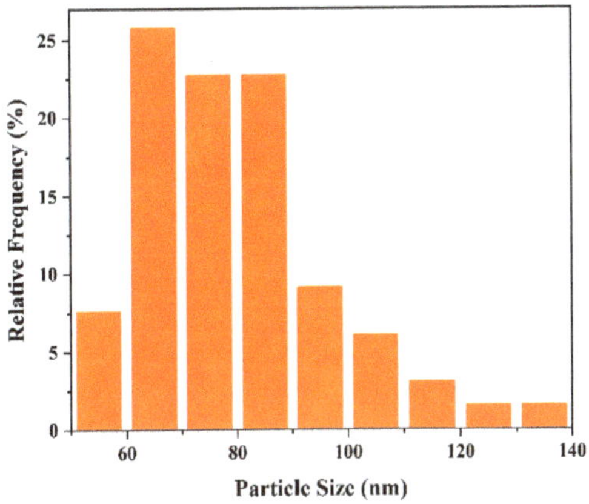

Figure 5. Particle size distribution of DST.

3.3. *Suitability*

3.3.1. Color Change

The color variation in the samples before and after treatment is shown in Figure 6. The color changes in the samples were mainly produced by ΔL^*, which was due to the application of the protective materials causing the stone surface to slightly darken, resulting in a slight decrease in L^*. However, the decrease in L^* was less than two. Both Δa^* and Δb^* were less than 0.5, indicating a small change in the coatings' red–green and yellow–blue coordinates and good transparency. Because of the good dispersion, no severe agglomeration of nanoparticles occurred to cover the original appearance of the stone surface, which maintained the excellent optical transparency of DTMS. There was no notable difference between adding only one nanoparticle or two nanoparticles at the same time, both of which had an ΔE of less than two, which did not change the color of the stones and met the requirement of $\Delta E < 5$ for materials used for stone conservation [42,43].

3.3.2. Water Vapor Permeability

Generally, the water vapor permeability of stone should remain unchanged after being treated with the ideal protective material; that is, the coating hinders the transport of liquid water but not that of water vapor. Water condensation underneath the protective layer leads to crispy alkali damage to the stone as a result of expansion and contraction under dry–wet cycles and contributes to the loss of adhesion and of the protective effects of the coating [31,44].

The water vapor permeability, δp, is shown in Table 2, and RP, the relative water vapor permeability, is the ratio of the water vapor permeability of the coated sample to the uncoated sandstone. The stone treated with DTMS and nanocomposite coatings maintained a relative water vapor permeability higher than 82% compared with the original sandstone, which meant they all had excellent water vapor permeability.

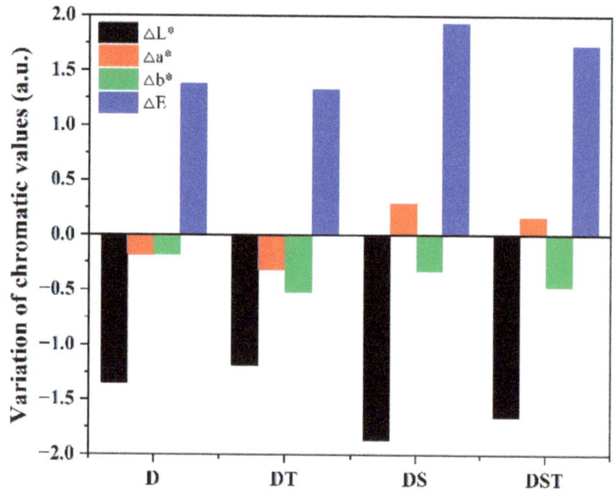

Figure 6. Variation in chromatic values before and after treatment.

3.4. *Durability*

3.4.1. Thermal Stability

Figure 7 shows the variation of the static contact angles (SCAs) of the samples during the thermal ageing test. Adding both TiO_2 and SiO_2 improved the heat resistance of DTMS, and DST was the most thermally stable.

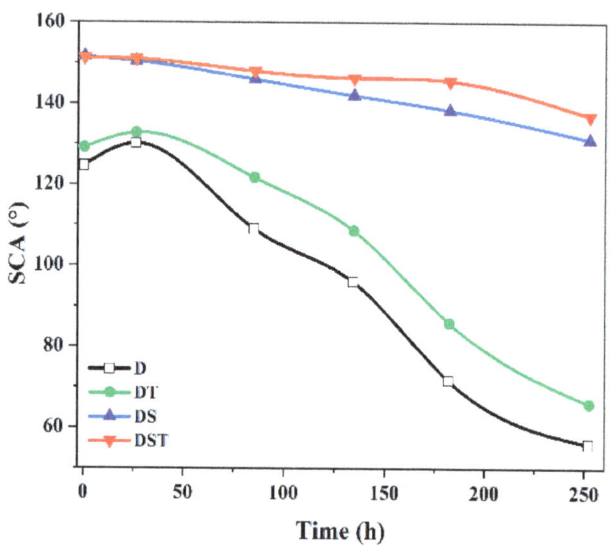

Figure 7. Variation in the static contact angles during the thermal ageing test.

Figure 8 shows the TG and DTG curves (the micro-quotient thermogravimetric curves) of the samples. As the coated sandstones were used as samples, the vast majority of which are mineral, there was minimal weight loss. The interval from 300 to 600 °C was chosen for our study because of minimal weight loss below 400 °C.

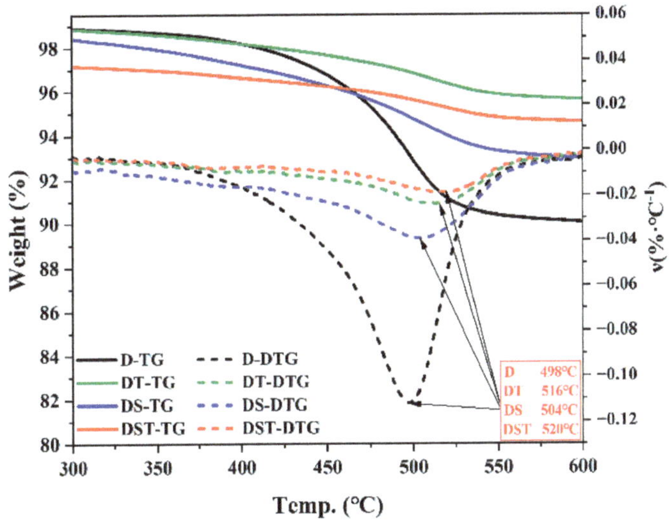

Figure 8. TG curves (solid lines) and DTG curves (dotted lines). Arrows point to maximum rates of weight loss. Values of temperature at the maximum rates of weight loss are provided in the lower right corner.

From the DTG curves in Figure 8, we found that the maximum weight loss rates occurred at 498 °C for DTMS, 516 °C for DT, 504 °C for DS, and 520 °C for DST. The rise in the maximum decomposition peak temperature values indicates an increase in thermal stability [45]. Therefore, the thermal stability of three modified coatings was improved compared with that of DTMS, where that of DT was improved more than that of DS. The two types of nanoparticles had a synergistic effect, so DST showed the highest thermal stability.

TiO_2 [46] and SiO_2 are both inorganic fire-retardant materials that can form passivation layers on a coating surface during thermal degradation. They can also act as insulators and/or mass transfer barriers to the volatile byproducts generated during thermal decomposition [47,48]. Additionally, increased thermal stability is related to the strong hydrogen bonding interactions of the nanoparticle and substrate, which impede the mobility of the molecular chains [49] and inhibit the attacks of free radicals on the main chain [45].

3.4.2. UV Shielding Performance

According to Rayleigh's theory, when the particle size is much smaller than the wavelength of the incident light, the intensity of the scattered light is inversely proportional to the fourth power of the incident wavelength ($I \propto \frac{1}{\lambda^4}$). Nanoparticles have strong UV light absorption and scattering effects because of their small particle size. With suitable dispersion, the addition of nanoparticles to coatings may provide shield from UV light [45,50] and improve the durability of coatings.

The variation in the contact angles of the samples during the UVA ageing test is shown in Figure 9. Compared with D, the UVA resistance of DT was lower. This may have been due to the photocatalytic activity of TiO_2 under UV light, which degraded DTMS under UVA irradiation [51], causing a decrease in its original hydrophobic property. Another explanation is that TiO_2 is photophilic [52] and hydrophilic under UV radiation, adsorbing invisible water films and causing a decrease in the contact angle [53].

The UVA resistance of DS was substantially stronger than that of D. After 144 h of UVA irradiation, the contact angle of D was 84.0°, whereas that of DS was 124.3°. The UVA resistance of DST was consistent with that of DS with no additional decrease, presumably due to the low concentration of TiO_2, which produced a negative effect that could be neglected compared with the UV shielding effect of SiO_2.

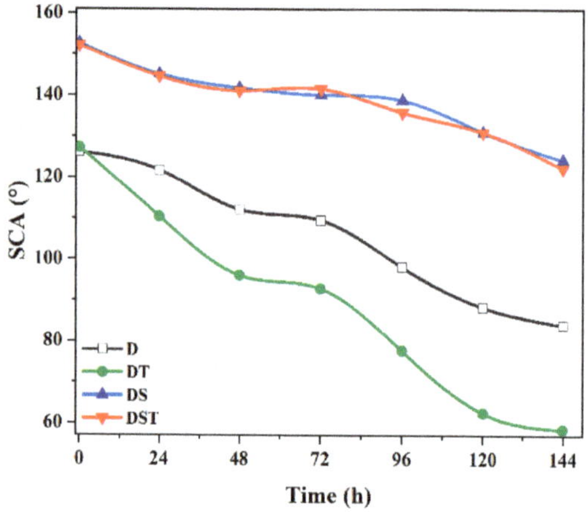

Figure 9. Variation in static contact angles during the UVA ageing test.

Figure 10 shows the transmittance–wavelength curves (solid lines) for different solutions in various UV regions: for the UVA, UVB, and UVC (200–400 nm) region and for the UVA (320–400 nm) region, the most common band in sunlight. The curves were integrated, which are shown as dashed lines in Figure 10. The UV resistance of the materials can be expressed by the shielding efficiency, which is calculated as follows [54]:

$$\eta_1 = \left(1 - \frac{\int_{200}^{400} T_{NP}(\lambda)d\lambda}{\int_{200}^{400} T_D(\lambda)d\lambda}\right) \times 100\% \tag{5}$$

$$\eta_2 = \left(1 - \frac{\int_{320}^{400} T_{NP}(\lambda)d\lambda}{\int_{320}^{400} T_D(\lambda)d\lambda}\right) \times 100\% \tag{6}$$

where η_1 and η_2 are the efficiency in shielding from 200–400 and 320–400 nm ultraviolet rays, respectively; $T(\lambda)$ is the average value of the spectral transmittance of the material; $d\lambda$ is the bandwidth; and λ is the wavelength. T_D is for the DTMS curve, and T_{NP} is DT, DS, or DST.

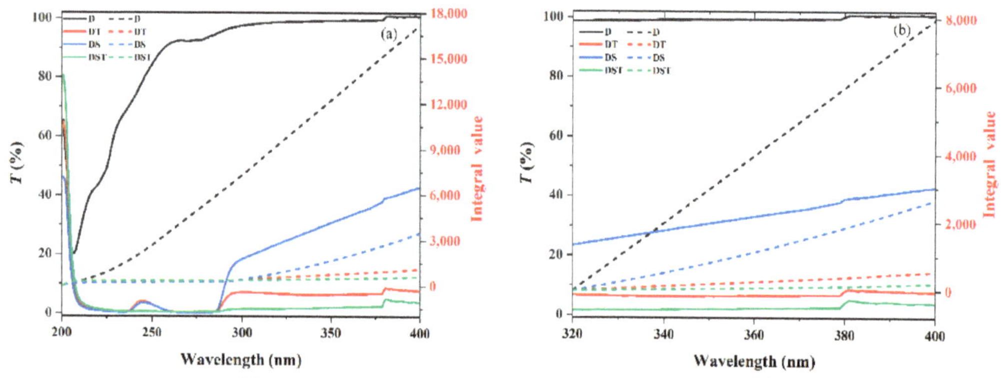

Figure 10. UV–Vis spectra of different groups of solutions in solid lines and integral values of the curves (right Y-axis) represented by dotted lines: (**a**) 200–400 nm; (**b**) 320–400 nm.

The shielding efficiency calculation is reported in Table 3. We found a strong UV shielding when SiO_2 and TiO_2 nanoparticles were added to DTMS. SiO_2 (0.5%) shielded nearly 80% of the UV in the 200–400 nm band and more than two-thirds of the most common UVA in sunlight. Low doses (0.01%) of TiO_2 shielded more than 93% of the whole wavelength band and the most common UVA bands. The mixture of the two nanoparticles synergistically shielded more than 95% and 97% of UV (200–400 nm) and UVA, respectively.

Table 3. UV shielding efficiency of the nanocomposite coatings.

Sample	$\int_{200}^{400} T(\lambda)d\lambda$	$\int_{320}^{400} T(\lambda)d\lambda$	η_1	η_2
D	17,292.80	7969.70	0	0
DT	1166.85	550.60	93.25	93.09
DS	3524.35	2636.80	79.62	66.91
DST	696.00	204.35	95.98	97.44

Combined with the result of UVA ageing, we found that TiO_2 had a high UV shielding efficiency but reduced the contact angle of coatings because the ultraviolet light shielded by TiO_2 does not safely radiate, which has a harmful effect on the coating [55,56]. Adding SiO_2 led to a safe radiation of the shielded ultraviolet, which prevented this adverse effect. DST still showed suitable UV shielding performance.

3.4.3. Chemical Stability

Figure 11 shows the changes in the contact angles of the samples immersed in solutions with different pH values for 24 h. The results showed that (1) The contact angle value of D and nanocomposites decreased the most (about 30°) under acidic conditions, and the heterogeneity increased (higher values of standard deviation), mainly because the sandstone contains calcite, which is dissolved under strongly acidic conditions, so the sandstone was structurally damaged, and many crystals precipitated on the surfaces (see the micromorphology of the samples in the Supplementary Materials Figure S2). The contact angle values of D, DT, and DS generally decreased by less than 5° after neutral water immersion, whereas that of DST remained unchanged. Under alkaline conditions, it only decreased by about 10°. (2) Adding nanoparticles to coatings, especially SiO_2, ensures the coating retains suitable hydrophobicity after immersion and improves its chemical stability.

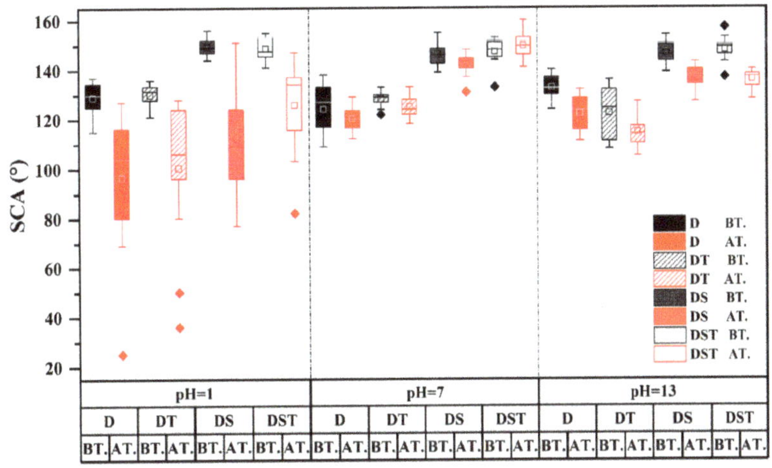

Figure 11. Changes in the contact angles of the samples immersed in solutions with different pH values for 24 h. BT, before treatment; AT, after treatment.

4. Conclusions

In this study, TiO_2 and SiO_2 nanoparticles were used to modify DTMS, a commonly used waterproof coating for stone artefacts, to fabricate nanocomposite coatings. We characterized the micromorphology, water-repellence, suitability for protecting stone heritage, and durability of the coatings. The results showed that the nanoparticles were well-dispersed in the coating, with 70% of the particle size of DST distributed in the range of 60 to 90 nm, with an average particle size of 81 nm, forming a rough nanoscale structure. The nanocomposite coatings had a minimal effect on the appearance of the stone, $\Delta E < 2$, and more than 82% of the water vapor permeability of the uncoated sandstone was retained, demonstrating suitability for stone artefact protection. The nanoparticles were added to DTMS: TiO_2 (0.01% *w/w*) failed to produce sufficient roughness and so had no impact on the contact angle of the coating but improved the thermal stability; SiO_2 (0.5% *w/w*) produced sufficient roughness, which made the coating superhydrophobic with a contact angle of more than 152° and improved the thermal, light, and chemical durability. When nano-TiO_2 and SiO_2 were simultaneously added, the nano-SiO_2 mitigated the effects of nano-TiO_2 accelerating UVA ageing and improved the original thermal stability, UVA shielding ability, and chemical stability of DTMS, resulting in a significant improvement in durability and DST showed the best overall performance.

The nano-SiO_2-TiO_2/DTMS, a superhydrophobic coating, fabricated in this study, has excellent properties, a low cost, and is simple to prepare and apply, showing promise as a material for protecting outdoor stone heritage.

Supplementary Materials: The following supporting information can be downloaded at https://www.mdpi.com/article/10.3390/coatings12101397/s1, Figure S1: XRD pattern of sandstone used in this study; Figure S2: Micromorphological images of the samples tested for chemical stability.

Author Contributions: M.P.: Investigation, Writing–original draft, Writing—review & editing. L.W.: Investigation, Supervision, Writing—review. L.G.: Investigation, Writing—review. J.G.: Investigation. L.Z.: Investigation. F.Y.: Investigation. Z.M.: Writing—review & editing. X.Z.: Supervision, Writing—review & editing. All authors have read and agreed to the published version of the manuscript.

Funding: This research was supported by the National Natural Science Foundation of China (52203126), the Key Research and Development Plan of Shaanxi Province (2019ZDLSF07-05, 2022SF-321), Projects of the Social Science Foundation of Shaanxi Province (2021G001), the Education Department of Shaanxi Province (20JZ092), National Social Science Fund of China (17XKG002), and School of Cultural Heritage, Northwest University (2020WYYCY-02), 111 Project (D18004).

Institutional Review Board Statement: Not applicable.

Informed Consent Statement: Not applicable.

Data Availability Statement: The datasets used and/or analyzed during the current study are available from the corresponding author on reasonable request.

Acknowledgments: We thank the Laboratory Technicians, Hongna Xing and Shuai Chang, of the Communion Center of Analysis & Measurement, School of Physics, Northwest University for the SEM and XRD analysis, respectively. Thanks to Yunlu Ma for her enlightening suggestions for data analysis.

Conflicts of Interest: The authors declare no conflict of interest.

References

1. Liu, R.; Zhang, B.; Zhang, H.; Shi, M. Deterioration of Yungang Grottoes: Diagnosis and research. *J. Cult. Herit.* **2011**, *12*, 494–499. [CrossRef]
2. Kaur, G.; Singh, S.; Kaur, P.; Garg, S.; Fareeduddin; Pandit, M.K.; Agrawal, P.; Acharya, K.; Ahuja, A. Vindhyan Sandstone: A Crowning Glory of Architectonic Heritage from India. *Geoheritage* **2019**, *11*, 1771–1783. [CrossRef]
3. Centauro, I.; Vitale, J.G.; Calandra, S.; Salvatici, T.; Natali, C.; Coppola, M.; Intrieri, E.; Garzonio, C.A. A Multidisciplinary Methodology for Technological Knowledge, Characterization and Diagnostics: Sandstone Facades in Florentine Architectural Heritage. *Appl. Sci.* **2022**, *12*, 4266. [CrossRef]

4. Rihosek, J.; Bruthans, J.; Masin, D.; Filippi, M.; Carling, G.T.; Schweigstillova, J. Gravity-induced stress as a factor reducing decay of sandstone monuments in Petra, Jordan. *J. Cult. Herit.* **2016**, *19*, 415–425. [CrossRef]
5. Migoń, P. Geoheritage and World Heritage Sites. *Geoheritage* **2018**, *13*, 237–249. [CrossRef]
6. Siegesmund, S.; Snethlage, R. *Stone in Architecture*; Springer: Berlin/ Heidelberg, Germany, 2011; p. 227. ISBN 978-3-642-14474-5.
7. Cao, Y.; Salvini, A.; Camaiti, M. One-step fabrication of robust and durable superamphiphobic, self-cleaning surface for outdoor and in situ application on building substrates. *J. Colloid Interface Sci.* **2021**, *591*, 239–252. [CrossRef]
8. Colangiuli, D.; Lettieri, M.; Masieri, M.; Calia, A. Field study in an urban environment of simultaneous self-cleaning and hydrophobic nanosized TiO_2-based coatings on stone for the protection of building surface. *Sci. Total Environ.* **2019**, *650*, 2919–2930. [CrossRef]
9. Pargoletti, E.; Motta, L.; Comite, V.; Fermo, P.; Cappelletti, G. The hydrophobicity modulation of glass and marble materials by different Si-based coatings. *Prog. Org. Coat.* **2019**, *136*, 105260. [CrossRef]
10. Serra, C.L.; Tulliani, J.-M.; Sangermano, M. An Acrylic Latex Filled with Zinc Oxide by Miniemulsion Polymerization as a Protective Coating for Stones. *Macromol. Mater. Eng.* **2014**, *299*, 1352–1361. [CrossRef]
11. Ruffolo, S.A.; La Russa, M.F. Nanostructured Coatings for Stone Protection: An Overview. *Front. Mater.* **2019**, *6*, 147. [CrossRef]
12. Cassie, A.B.D.; Baxter, S. Wettability of porous surfaces. *Trans. Faraday Soc.* **1944**, *40*, 546. [CrossRef]
13. Xue, C.; Jia, S.; Zhang, J.; Ma, J. Large-area fabrication of superhydrophobic surfaces for practical applications: An overview. *Sci. Technol. Adv. Mater.* **2010**, *11*, 33002. [CrossRef] [PubMed]
14. Saji, V.S. Electrophoretic-deposited Superhydrophobic Coatings. *Chem. Asian J.* **2021**, *16*, 474–491. [CrossRef]
15. Li, M.; Yang, Y.-Q.; Liu, L.; Hu, J.; Zhang, J. Electro-assisted preparation of dodecyltrimethoxysilane/TiO_2 composite films for corrosion protection of AA2024-T3 (aluminum alloy). *Electrochim. Acta* **2010**, *55*, 3008–3014. [CrossRef]
16. Zhang, S.; Cao, D.; Xu, L.; Tang, J.; Meng, R.; Li, H. Corrosion resistance of a superhydrophobic dodecyltrimethoxysilane coating on magnesium hydroxide-pretreated magnesium alloy AZ31 by electrodeposition. *Colloids Surf. A* **2021**, *625*, 126914. [CrossRef]
17. Xu, B.; Cai, Z. Fabrication of a superhydrophobic ZnO nanorod array film on cotton fabrics via a wet chemical route and hydrophobic modification. *Appl. Surf. Sci.* **2008**, *254*, 5899–5904. [CrossRef]
18. Zheng, L.; Chen, Y. Research on the Restoration Techniques and Prevention Strategies of Historic Buildings in Macao. *IOP Conf. Ser. Earth Environ. Sci.* **2021**, *783*, 12111. [CrossRef]
19. Zheng, L.; Chen, Y. Problems and countermeasures in the restoration project of cultural relics of Macao Chong Sai Pharmacy. *E3S Web Conf.* **2021**, *284*, 5005. [CrossRef]
20. Ling, H.; Maiqian, N.; Guozheng, L. Preparation and feasibility analysis of fluoropolymer to the sandstone protection. *Prog. Org. Coat.* **2008**, *62*, 206–213. [CrossRef]
21. Zhen, G. Application of WD-10 in the surface protection of the lithoid cultural relic. *New Chem. Mater.* **2001**, 48–50. Available online: https://www.cnki.com.cn/Article/CJFDTOTAL-HGXC200109011.htm (accessed on 29 August 2022).
22. Wang, H.; Feng, N.; Song, D. Study of conservation method for the weathering of Goguryeo stone heritage. *Relics Museol.* **2010**, 76–81. Available online: https://www.cnki.com.cn/Article/CJFDTOTAL-WEBO201006017.htm (accessed on 29 August 2022).
23. Wang, L.; Li, Y.; Zhao, X. Modification of WD-10, a waterproof material for stone relics, with nano TiO_2. *Fine Chem.* **2015**, *32*, 250–253. Available online: https://www.cnki.com.cn/Article/CJFDTOTAL-JXHG201503003.htm (accessed on 29 August 2022). [CrossRef]
24. Baglioni, P.; Carretti, E.; Chelazzi, D. Nanomaterials in art conservation. *Nat. Nanotechnol.* **2015**, *10*, 287–290. [CrossRef] [PubMed]
25. Franco-Castillo, I.; Hierro, L.; de La Fuente, J.M.; Seral-Ascaso, A.; Mitchell, S.G. Perspectives for antimicrobial nanomaterials in cultural heritage conservation. *Chem* **2021**, *7*, 629–669. [CrossRef]
26. Honarvar Nazari, M.; Zhang, Y.; Mahmoodi, A.; Xu, G.; Yu, J.; Wu, J.; Shi, X. Nanocomposite organic coatings for corrosion protection of metals: A review of recent advances. *Prog. Org. Coat.* **2022**, *162*, 106573. [CrossRef]
27. Dalawai, S.P.; Saad Aly, M.A.; Latthe, S.S.; Xing, R.; Sutar, R.S.; Nagappan, S.; Ha, C.; Kumar Sadasivuni, K.; Liu, S. Recent Advances in durability of superhydrophobic self-cleaning technology: A critical review. *Prog. Org. Coat.* **2020**, *138*, 105381. [CrossRef]
28. Frigione, M.; Lettieri, M. Novel attribute of organic–inorganic hybrid coatings for protection and preservation of materials (stone and wood) belonging to cultural heritage. *Coatings* **2018**, *8*, 319. [CrossRef]
29. Facio, D.S.; Mosquera, M.J. Simple strategy for producing superhydrophobic nanocomposite coatings in situ on a building substrate. *ACS Appl. Mater. Interfaces* **2013**, *5*, 7517–7526. [CrossRef]
30. Manoudis, P.N.; Karapanagiotis, I.; Tsakalof, A.; Zuburtikudis, I.; Kolinkeová, B.; Panayiotou, C. Superhydrophobic films for the protection of outdoor cultural heritage assets. *Appl. Phys. A* **2009**, *97*, 351–360. [CrossRef]
31. Caponetti, E.; Ciaramitaro, V.; Armetta, F.; Renda, V.; Ercoli, L.; Saladino, M.L. Effectiveness of some protective and self-cleaning treatments: A challenge for the conservation of temple G stone in Selinunte. *Prog. Org. Coat.* **2021**, *151*, 106020. [CrossRef]
32. Chatzigrigoriou, A.; Karapanagiotis, I.; Poulios, I. Superhydrophobic coatings based on siloxane resin and calcium hydroxide nanoparticles for marble protection. *Coatings* **2020**, *10*, 334. [CrossRef]
33. Ditaranto, N.; Van der Werf, I.D.; Picca, R.A.; Sportelli, M.C.; Giannossa, L.C.; Bonerba, E.; Tantillo, G.; Sabbatini, L. Characterization and behaviour of ZnO-based nanocomposites designed for the control of biodeterioration of patrimonial stoneworks. *New J. Chem.* **2015**, *39*, 6836–6843. [CrossRef]

34. Essa, A.M.; Khallaf, M.K. Biological nanosilver particles for the protection of archaeological stones against microbial colonization. *Int. Biodeterior. Biodegrad.* **2014**, *94*, 31–37. [CrossRef]
35. Karapanagiotis, I.; Manoudis, P.N. Superhydrophobic and superamphiphobic materials for the conservation of natural stone: An overview. *Constr. Build. Mater.* **2022**, *320*, 126175. [CrossRef]
36. La Russa, M.F.; Rovella, N.; Alvarez de Buergo, M.; Belfiore, C.M.; Pezzino, A.; Crisci, G.M.; Ruffolo, S.A. Nano-TiO_2 coatings for cultural heritage protection: The role of the binder on hydrophobic and self-cleaning efficacy. *Prog. Org. Coat.* **2016**, *91*, 1–8. [CrossRef]
37. *UNI EN 15801:2010*; Conservation of Cultural Property-Test Methods—Determination of Water Absorption by Capillarity. UNI Ente Nazionale Italiano di Unificazione: Milan, Italy, 2010.
38. Camaiti, M.; Brizi, L.; Bortolotti, V.; Papacchini, A.; Salvini, A.; Fantazzini, P. An Environmental Friendly Fluorinated Oligoamide for Producing Nonwetting Coatings with High Performance on Porous Surfaces. *ACS Appl. Mater. Interfaces* **2017**, *9*, 37279–37288. [CrossRef] [PubMed]
39. *UNI EN 15803:2010*; Conservation of Cultural Property-Test Methods—Determination of Water Vapour Permeability. UNI Ente Nazionale Italiano di Unificazione: Milan, Italy, 2010.
40. Karapanagiotis, I.; Manoudis, P.N.; Savva, A.; Panayiotou, C. Superhydrophobic polymer-particle composite films produced using various particle sizes. *Surf. Interface Anal.* **2012**, *44*, 870–875. [CrossRef]
41. Manoudis, P.N.; Karapanagiotis, I. Modification of the wettability of polymer surfaces using nanoparticles. *Prog. Org. Coat.* **2014**, *77*, 331–338. [CrossRef]
42. La Russa, M.F.; Ruffolo, S.A.; Rovella, N.; Belfiore, C.M.; Palermo, A.M.; Guzzi, M.T.; Crisci, G.M. Multifunctional TiO_2 coatings for Cultural Heritage. *Prog. Org. Coat.* **2012**, *74*, 186–191. [CrossRef]
43. García, O.; Malaga, K. Definition of the procedure to determine the suitability and durability of an anti-graffiti product for application on cultural heritage porous materials. *J. Cult. Herit.* **2012**, *13*, 77–82. [CrossRef]
44. Lettieri, M.; Masieri, M.; Frigione, M. Durability to simulated bird guano of nano-filled oleo/hydrophobic coatings for the protection of stone materials. *Prog. Org. Coat.* **2020**, *148*, 105900. [CrossRef]
45. Loste, J.; Lopez-Cuesta, J.-M.; Billon, L.; Garay, H.; Save, M. Transparent polymer nanocomposites: An overview on their synthesis and advanced properties. *Prog. Polym. Sci.* **2019**, *89*, 133–158. [CrossRef]
46. Zhou, L.; Fu, Y. Flame-Retardant Wood Composites Based on Immobilizing with Chitosan/Sodium Phytate/Nano-TiO_2-ZnO Coatings via Layer-by-Layer Self-Assembly. *Coatings* **2020**, *10*, 296. [CrossRef]
47. Cai, Y.; Ke, H.; Dong, J.; Wei, Q.; Lin, J.; Zhao, Y.; Song, L.; Hu, Y.; Huang, F.; Gao, W.; et al. Effects of nano-SiO_2 on morphology, thermal energy storage, thermal stability, and combustion properties of electrospun lauric acid/PET ultrafine composite fibers as form-stable phase change materials. *Appl. Energy* **2011**, *88*, 2106–2112. [CrossRef]
48. Díez-Pascual, A.M.; Xu, C.; Luque, R. Development and characterization of novel poly(ether ether ketone)/ZnO bionanocomposites. *J. Mater. Chem. B* **2014**, *2*, 3065–3078. [CrossRef] [PubMed]
49. Díez-Pascual, A.M.; Díez-Vicente, A.L. Nano-TiO_2 reinforced PEEK/PEI blends as biomaterials for load-bearing implant applications. *ACS Appl. Mater. Interfaces* **2015**, *7*, 5561–5573. [CrossRef] [PubMed]
50. Liu, H; Hu, D.; Chen, X.; Ma, W. Surface engineering of nanoparticles for highly efficient UV -shielding composites. *Polym. Adv. Technol.* **2021**, *32*, 6–16. [CrossRef]
51. Li, W.; Lin, J.; Zhao, Y.; Pan, Z. The Adverse Effects of TiO_2 Photocatalycity on Paraloid B72 Hybrid Stone Relics Protective Coating Aging Behaviors under UV Irradiation. *Polymers* **2021**, *13*, 262. [CrossRef]
52. Wang, R.; Hashimoto, K.; Fujishima, A.; Chikuni, M.; Kojima, E.; Kitamura, A.; Shimohigoshi, M.; Watanabe, T. Light-induced amphiphilic surfaces. *Nature* **1997**, *388*, 431–432. [CrossRef]
53. Colangiuli, D.; Calia, A.; Bianco, N. Novel multifunctional coatings with photocatalytic and hydrophobic properties for the preservation of the stone building heritage. *Constr. Build. Mater.* **2015**, *93*, 189–196. [CrossRef]
54. Niu, X.; Liu, Y.; Fang, G.; Huang, C.; Rojas, O.J.; Pan, H. Highly Transparent, Strong, and Flexible Films with Modified Cellulose Nanofiber Bearing UV Shielding Property. *Biomacromolecules* **2018**, *19*, 4565–4575. [CrossRef]
55. Miliani, C.; Monico, L.; Melo, M.J.; Fantacci, S.; Angelin, E.M.; Romani, A.; Janssens, K. Photochemistry of Artists' Dyes and Pigments: Towards Better Understanding and Prevention of Colour Change in Works of Art. *Angew. Chem. Int. Edit.* **2018**, *57*, 7324–7334. [CrossRef]
56. Samain, L.; Gilbert, B.; Grandjean, F.; Long, G.J.; Strivay, D. Redox reactions in Prussian blue containing paint layers as a result of light exposure. *J. Anal. At. Spectrom.* **2013**, *28*, 524. [CrossRef]

coatings

MDPI

Article

Preliminary Investigation of Sequential Application of Different Calcium Oxalate Solutions for Carbonate Rock Conservation

Jianrui Zha [1], Yaoqi Gu [1], Shuya Wei [1,*], Huarui Han [2], Ankun Wang [3] and Qinglin Ma [4,*]

1 Institute of Cultural Heritage and History of Science and Technology, University of Science and Technology Beijing, Beijing 100083, China
2 Institute of Archaeology, Chinese Academy of Social Sciences, Beijing 100101, China
3 Henan Provincial Institute for the Conservation of Architectural Heritage, Zhengzhou 450002, China
4 International Joint Research Laboratory of Environmental and Social Archaeology, Shandong University, Qingdao 266237, China
* Correspondence: sywei66@hotmail.com (S.W.); qinglinma226@126.com (Q.M.); Tel.: +86-18701318128 (S.W.)

Abstract: Being inspired by nature, a series of experiments was carried out to deposit a calcium oxalate layer on the surface of the stone by the reaction between carbonate rock and oxalate salt. To increase the anti-dissolution properties of the calcium oxalate layer, the use of mixed oxalate solution has been proposed in the literature by two main routes: (1) adding acid agent to ammonium oxalate, which has the advantage of changing the particle structure and reducing layer porosity, and (2) using neutral methyl oxalate solution, which has the advantage of surface coverage due to slowly hydrolysis. In this study, we investigated the sequential application of ammonium oxalate, methyl oxalate, neutral mixed, and calcium acetate acid mixed solution. With this method, calcium carbonate and calcium oxalate solution can react inside the stone to reinforce it. The protective film's coverage area can then be increased using dimethyl oxalate neutral mixed solution, and the crystal morphology can be modified with calcium oxalate acid mixed solution. The anti-dissolution properties of the coating were investigated using both a custom-designed apparatus and a selective outdoor environment. The coating displayed good acid resistance properties at pH 2–4. After one year of exposure, the coating is firmly bonded with the stone.

Keywords: carbonate rock; sequential application; whewellite

Citation: Zha, J.; Gu, Y.; Wei, S.; Han, H.; Wang, A.; Ma, Q. Preliminary Investigation of Sequential Application of Different Calcium Oxalate Solutions for Carbonate Rock Conservation. *Coatings* **2022**, *12*, 1412. https://doi.org/10.3390/coatings12101412

Academic Editor: Enrico Quagliarini

Received: 12 August 2022
Accepted: 22 September 2022
Published: 27 September 2022

1. Introduction

In China, a large part of stone cultural heritage has suffered from chemical dissolution, and the process is accelerated in polluted areas by the presence of nitrogen and sulfur oxides, which have a synergistically accelerating effect [1]. In order to reduce the effect of chemical weathering, coating interventions can be used, especially with outdoor stone artifacts with a high open porosity [2]. Various organic and inorganic chemicals have been proposed for the conservation of stone cultural heritage surfaces in order to improve the substrate water-repellent and anti-dissolution properties [3,4]. However, organic polymer materials were demonstrated to be less compatible with the inorganic stone cultural heritages, particularly carbonate rock [5]. In certain cases, inorganic protection coatings are more compatible with carbonate rock. Some hydroxides, such as calcium hydroxide and magnesium hydroxide, have already been utilized as consolidant because they can react with atmospheric carbon dioxide and form a compatible inorganic coating on the stone surface [6–10]. Additionally, inorganic coatings such as calcium oxalate and calcium phosphate have been confirmed to effectively inhibit the dissolution related phenomenon of stone monuments and temples in Europe [11,12]. Because it can be removed, calcium oxalate is receiving much more attention in the field of heritage protection than calcium phosphate [13]. The idea of producing artificial calcium oxalate as a protective layer to provide carbonate rock with an acid-resistant surface originated in the nineteenth century and numerous methods have been developed [14]. It is notable that the calcium oxalate layer can be obtained by reacting different kinds of solution with calcium ions originating from the substrate [15]. The treatment consists of applying a poultice of ammonium oxalate, dimethyl oxalate, and oxalic acid to carbonate rock surface, with the purpose of forming a calcium oxalate monohydrate layer [16] (Equations (1)–(3)).

$$Ca^{2+} + C_2O_4{}^{2-} + H_2O \rightarrow CaC_2O_4 \cdot H_2O, \quad (1)$$

$$CH_3OOCCOOCH_3 + H_2O \rightarrow H_2C_2O_4 + CH_3CH_2OH, \quad (2)$$

$$H_2C_2O_4 + CaCO_3 \rightarrow CaC_2O_4 \cdot H_2O + CO_2, \quad (3)$$

Notably, unreacted solutions, other calcium oxalate phases (calcium oxalate dihydrate and metastable crystalline CaOx phases), and reaction by-products (magnesium oxalate phase), all of which have the potential to influence the coating properties, remain at the end of the treatment, as confirmed by multiple techniques in previous studies [17,18]. According to researchers, a major portion of this problems were related to heterogeneous substrates. For example, carbonate rock comprises of calcite and dolomite. When dolomite contributes Ca^{2+} and Mg^{2+} ions to the coating-formation process, calcium and magnesium oxalate rims are produced, respectively. The high solubility of magnesium oxalate in water, and the restricted spatial distribution of the magnesium oxalate phase influence the coating coverage [19]. Therefore, a variety of oxalate salt solution combinations were investigated in an effort to prevent the described drawback [20].

The advantages of the oxalate salt mixed solution were initially proposed in the modification of crystal structure and afterwards assessed in the extension of the coating coverage area [21]. Current research on marble and limestone samples has shown that acid solution treatments affect whewellite crystal preferred orientations. This leads to a more homogenous crystal structure, which reduces porosity [22]. Moreover, studies on marble found that adding neutral diethyl oxalate solution can dramatically increase surface coverage while lowering cracking and porosity. This is because the molecule slowly hydrolyzes to minimize ionic interaction throughout the process of dissolution and participation [23]. This, in turn, increases the resistance of the samples against acid dissolution. If the solutions described above could work together simultaneously, the protective effect of the calcium oxalate layer would be enhanced. However, in the absence of acid, the hydrolysis rate of methyl oxalate increases, and the resulting product oxalic acid loses the advantage effect [24]. This remains a big challenge.

The calcium phosphate layer deposition experiment may provide inspiration for conservators seeking solutions to their issues. In a previous study, it was discovered that application of a limewater poultice after treatment with diammonium hydrogen phosphate increased calcium phosphate formation and removed the unreacted agent [25]. However, when ammonium oxalate and diammonium hydrogen phosphate are applied sequentially, a patchy layer develops [26]. This is due to the fact that whewellite is formed when ammonium oxalate reacts with calcite, which reduces the stone's porosity. As a result, when the diammonium hydrogen phosphate solution is applied, there are no more calcium ions on the surface to form calcium phosphate, and deep penetration of the diammonium hydrogen phosphate solution into the substrate is inhibited. So, layer deposition can be improved by the sequential application of calcium gel or micro-grout solutions.

Therefore, in this research the sequential application of ammonium oxalate, methyl oxalate neutral mixed and calcium acetate acid mixed solution on dolomite marble was investigated. To evaluate the effect, firstly, the chemical composition and structure morphology of the deposition layer were characterized. The resistance to dissolution was assessed by being exposed to specific outdoor environments as well as a custom apparatus that could simulate rain, as in the references [26]. Through the above dissolution resistant experiment, it is possible to find the effects of the calcium oxalate layer, which provides direction for the sequential application method. This research concentrates on the study of coating densification and crystal regulation, while other research gives more attention to the preparation method, application environment, and performance evaluation [6,27].

2. Materials and Methods

2.1. Coating Formulation

For the coating preparation, ammonium oxalate ($(NH_4)_2C_2O_4$), calcium acetate ($Ca(CH_3COO)_2\cdot H_2O$), oxalic acid ($H_2C_2O_4$), nitric acid (HNO_3), dimethyl oxalate ($C_4H_6O_4$), sodium hydroxide (NaOH), and ethanol (C_2H_5OH) were used. All of the chemicals were acquired from Aladdin-reagent Co., Ltd. (Shanghai, China) and without further purification. Three distinct solutions were obtained as previous report: 0.1 mol/L ammonium oxalate solution ($(NH_4)_2C_2O_4 + H_2O$), 0.02 mol/L Calcium oxalate solution A ($Ca(CH_3COO)_2\cdot H_2O$ + $C_4H_6O_4$+ NaOH + H_2O + C_2H_5OH), and 0.025 mol/L Calcium oxalate solution B ($Ca(CH_3COO)_2\cdot H_2O + H_2C_2O_4 + HNO_3 + H_2O$). To obtain the best results when applied to outdoor stone, the entire drying period of the solution was set at more than 12 h.

As depicted in Figure 1, obtained solutions were sequentially applied to the selected substrate: one application of ammonium oxalate solution, four applications of Calcium oxalate solution A, and one application of Calcium oxalate solution B. Coatings were deposition on a modest substrate (25 mm × 25 mm × 10 mm) for laboratory analysis, and on a natural stone for outdoor exposure. The coating was cured by drying for 24 h at room temperature until a consistent mass was observed.

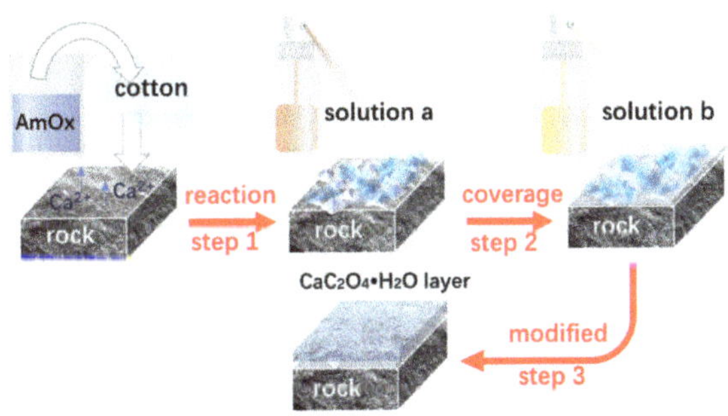

Figure 1. Scheme illustrating the treatment process of each sample.

2.2. Stone Substrate

Two kinds of calcareous stone substrates were selected to evaluate the composition, morphology, and performance of the coating. Modest marble substrate (Angera stone) with dimensions of 25 mm × 25 mm × 10 mm was obtained from the local market and polished with SiC paper (180#). Nature stone (green-white stone) was collected from the Fangshan district, where a large quantity of stone relics had been unearthed. Both stones were ultrasonic cleaned with deionized water and baked for 24 h at 110 °C in order to remove the dust on the surface.

2.3. Characterization Techniques

X-ray diffraction (XRD, Bruker D8 Advance, Karlsruhe, Germany) analysis was used to analyze the coated stone and acid rain attacked samples. The X-rays were generated with Cu Kα radiation at 45 kV and 200 mA in a scanning range of 10–70(2θ) with a scanning speed of 1°/min and step size of 0.02°. Before measurements, the samples were cut to meet the requirements of the analysis.

Raman spectrum were performed by using a Horiba Scientific LabRAM HR confocal Raman spectrometer (Horiba Scientific, Lille, France). The coated stone, the acid rain attacked samples and the outdoor exposure samples were excited with 785 nm laser, using the 50× objective lens with 1 μm spatial resolution. In order to improve the signal to noise ratio, the acquisition time was fixed at 60 s with 1 accumulation. The analyses performed on surface sections placed on the motorized X-Y microscope stage. Raman spectra were recorded automatically along an area. A great advantage of this method is its high spatial resolution (4–5 μm). In addition, a visible-light microscope allows characterized of a selected area. So, the distribution of the phases was obtained by area maps.

Fourier transform infrared spectra were collected on FTIR spectrophotometer, Thermo Fisher Nicolet (Waltham, MA, USA) is 5, in the range of 500–4000 cm^{-1}, using resolution 4 and 16 scan. The coated marble substrate was done using attenuated total reflection (ATR) accessory.

Scanning electron microscope (SEM) images were obtained using Tescan Vega3 (Brno, Czech Republic) from surface of coated stone, the acid rain attacked samples, the outdoor exposure samples, and cross-sections of coated stone as well as the outdoor exposure samples. All of the samples were sputter-coated with gold, and observed at an accelerating voltage of 20 kV.

The surface roughness of coated stone was imaged in the tapping mode in air using Bruker Nano Dimension Edge AFM (Bruker, Billerica, MA, USA). Samples were calculated at controlled room temperature (22 °C).

The acid resistance of the specimens was evaluated by simulating the rain dissolution process learn from Graziani reference [26]. Acid solution composed of HNO_3 and H_2SO_4 (1:1, pH = 1, 2, 3, 4) was dripped onto the coated or uncoated marble samples. Furthermore, the coated green-white stone began to be exposed outside in February of 2021. After one year of exposure, the environmental degradation of the sample was evaluated. Table 1 summarizes the characteristics of the preceding samples.

Table 1. Characteristics of the samples.

Sample	Type	Test Description	Analysis Description
1	coated stone	according to analysis requirement	XRD, Raman, FTIR, AFM, SEM (surface and cross section)
2	stone-1	simulate acid rain dissolution experiment at pH 1	XRD, Raman, SEM (surface)
3	stone-2	simulate acid rain dissolution experiment at pH 2	XRD, Raman, SEM (surface)
4	stone-3	simulate acid rain dissolution experiment at pH 3	XRD, Raman, SEM (surface)
5	stone-4	simulate acid rain dissolution experiment at pH 4	XRD, Raman, SEM (surface)
6	coated stone-1	simulate acid rain dissolution experiment at pH 1	XRD, Raman, SEM (surface)
7	coated stone-2	simulate acid rain dissolution experiment at pH 2	XRD, Raman, SEM (surface)
8	coated stone-3	simulate acid rain dissolution experiment at pH 3	XRD, Raman, SEM (surface)
9	coated stone-4	simulate acid rain dissolution experiment at pH 4	XRD, Raman, SEM (surface)
10	coated green-white stone	exposure to outside environment (Beijing huairou district)	Digital image, Raman mapping, SEM (surface section)

3. Results and Discussion

3.1. Characterization of the Film

As previously stated, it is well known that the application of ammonium oxalate solution can form a $CaC_2O_4{\cdot}H_2O$ coating, thereby inhibiting the degradation of the stones [28]. In order to determine the influence of combined treatment with ammonium oxalate solution, calcium oxalate solution A, and calcium oxalate solution B, the coated marble stones' composition was thoroughly investigated.

First of all, XRD studies of the crystalline phases (Figure 2a) reveals the corresponding Bragg peaks for $CaMg(CO_3)_2$ and $CaC_2O_4{\cdot}H_2O$ in the coated stone. It is not surprising that weddellite and glushinskite are missing under the combined treatment. In fact, a number of studies discovered that only whewellite precipitated at pH < 5, whereas glushinskite was absent from the treated dolomite marble due to low amount of this phases in the reaction rim [29]. Moreover, the Raman spectra (Figure 2b) of the coated stone reveals that the peaks of whewellite at 140, 192, 246, 501, 597, 725, 895, 1462, 1489 cm^{-1} were almost not shifted to the other wavenumbers [30]. This result may be attributed to the absence of amorphous magnesium oxalate, which is usually detected in dolomite treated with ammonium oxalate solution. After the combined use of different calcium oxalate solution and subsequent washing, the possibility amorphous magnesium oxalate was washed out. Moreover, FTIR reveals a weak Fe–O bond at 585 cm^{-1}, with the exception of the whewellite group (Table 2). The associated mineral still present thanks to the lower impact of ammonium oxalate solution and neutral pH value of the calcium oxalate solution A [31].

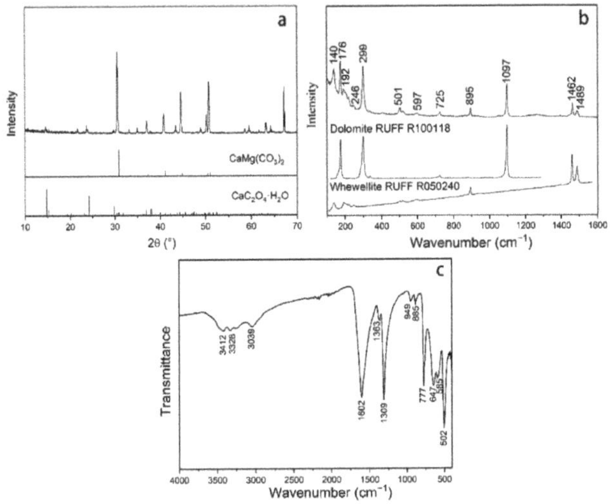

Figure 2. Composition of treated samples. (**a**) XRD, (**b**) Raman, (**c**) FTIR.

Table 2. 2θ degree of test samples.

Samples	2θ (°)	hkl
Dolomite (36–0426)	30.938	104
Coated Stone-1	30.937	104
Coated Stone-2	30.960	104
Coated Stone-3	30.982	104
Coated Stone-4	30.978	104
Stone-1	30.880	104
Stone-2	30.940	104
Stone-3	31.000	104
Stone-4	30.940	104

In order to investigate the morphology of the newly formed crystals, the same portions of the treated samples analyzed by Raman spectra were observed by scanning electron microscope. The sample surface is completely coated with calcium oxalate as observed in Figure 3a. The crystals of whewellite are grouped in agglomerates with an appreciable tabular structure. The sizes of many crystals are around 1 μm. Interestingly, a homogenous calcium oxalate layer without distinct layer structure is observed on the dolomite marble surface after combined treatment (Figure 3b); also, etch pits are not visible on the substrate surface. Furthermore, the inner layer between the unreacted substrate and the coating has a comparable texture to the surface layer. It is unclear if this is the penetration result of the solution A or solution B, or a consequence of the combined deposition process. These results are consistent with AFM morphological analysis (Figure 3c). The surfaces of stone are covered with tabular clusters of uniform size and low standard deviations for both particle species. Moreover, pseudocolor pictures indicate the depth field of surface features, in which bright color represents the elevated features, and darker colors represent deeper ones. Variation in the topographical image of the surface indicates the roughness of the whole sample. However, considering the roughness structure of dolomite basement, these roughness values are relatively lower than those of traditional treatment, as shown in Figure S1 (see Supplementary Materials).

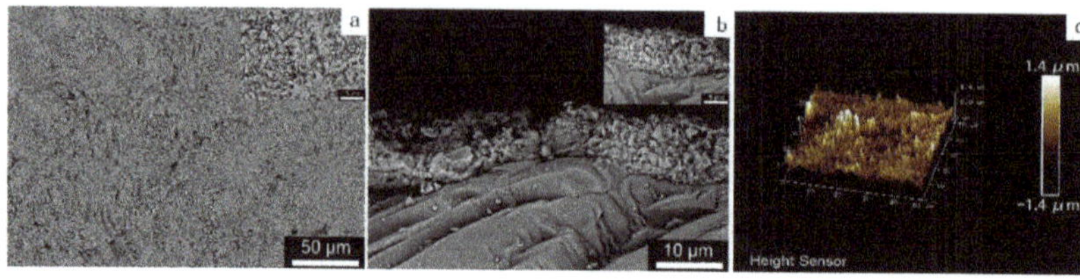

Figure 3. Morphology of treated samples. (**a**) SEM-surface section, (**b**) SEM-cross section, (**c**) AFM-surface section.

3.2. Protective Properties of the Film

Stone and coated stone were subjected to a dripping test at pH 1, 2, 3, 4 to evaluate their resistance to simulate acid rain dissolution. Lower pH values were selected in order to reproduce severe conditions and observe the specimens' deterioration. In fact, the dissolution process of calcium oxalate is related to ionic strength, protonation reactions, and the production of soluble complex [32]. The laboratory experiments focus on the dissolution mechanism, which is the most important part of the dissolution process. Considering the changeable outdoor environment, a field aging test should also be conducted on the coated samples to evaluate the protective properties.

Regarding the dissolution appearance of stone, the XRD pattern (Figure 4a) exhibits the dolomite diffraction peaks around 30.9°, 35.3°, 37.4°, 41.1°, 44.9°, 51.1°, 58.9°, and 67.4°, corresponding to (104), (015), (110), (113), (202), (116), (211), and (300) preferred orientations, respectively. These values are consistent with the international Center of Diffraction Data (ICDD) card number (36-0426). The strongest peak occurs at 2θ~30.9°, which is referred to (104) plane. The position of the peaks shifted after simulating acid rain dissolution, and the coated stone was altered at pH 1–2, leading to the conclusion that the calcium oxalate layer was destroyed during the dissolution process. In addition, it can be noticed that 2θ for stone changes irregularly, and maybe other impure constituents affects the result. The absence of other aging products suggests that the $CaSO_4$ or $Ca(NO)_3$ crystals are easily washed out with runoff solution. In Figure 4b, we compare the Raman spectra of stone and coated stone treated with acid rain. The main peak of dolomite is not shifted, and two samples (coated stone-3 and coated stone-4) exhibit prominent whewellite peaks when the pH value of simulated acid solution is 3 and 4. An obvious rise in dolomite peak width (FWHM) at higher wave numbers is detected at Coated stone-2, Coated stone-1, and stone-1. The feature is the result of the dissolution of other crystals. Therefore, the stone and the coated stone undergo two different dissolution processes.

In order to better characterize the sample dissolution process, the surface morphologies of all samples after simulated acid rain dissolution are shown in Figure 5. On the stone surfaces, cracks of varying degrees emerge, signifying the different levels of degradation correlating to the pH value of acid solution. In comparison, coated stone exhibits vast changes after simulated acid rain dissolution. The surface denudation of the sample treated with pH 1 solution confirms that the sample was severely damaged. On the surface of the other treated sample, only little holes are presented, implying stronger dissolution resistance. It is possible that the uniform calcium oxalate layer inhibited the dissolution process on the stone's surface. The detachment of tightly bonded coating promotes the destruction of substrate. In order to better reflect the performance of this treatment, we compared the results with our previous studies using part components of the chemicals, including using ammonium oxalate individually or together with solution A or solution B. All the corresponding SEM images are shown in Figures S2 and S3 (see Supplementary Materials). This comparison can prove that the coating formed through this treatment is more uniform.

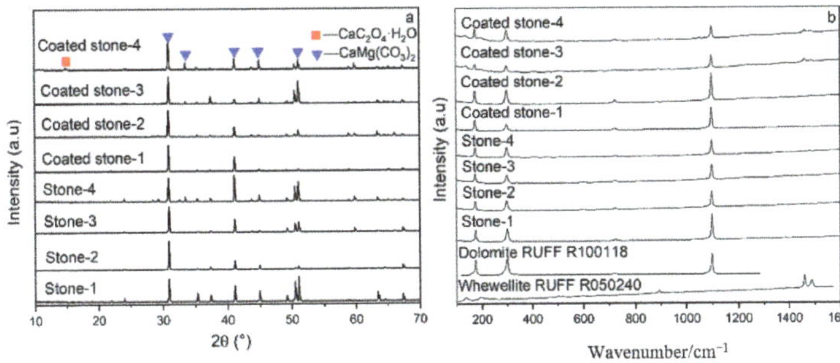

Figure 4. Composition of test samples after treatment with simulated acid rain at different pH values. (**a**) XRD, (**b**) Raman.

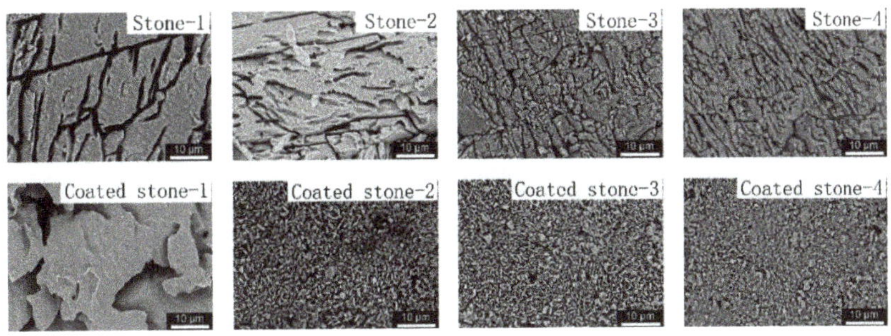

Figure 5. SEM microscope of test samples after treatment with simulated acid rain at different pH values.

As mentioned in the introduction, the deterioration process of stone is influenced by the chemical, physical, and biological factors. Field exposure tests give the possibility to evaluate the environmental impact on the applied conservation actions. So, treated samples were subjected to exposure tests in typical environmental at a house estate in Beijing's Huairou District. There is no obvious change to the surface of green-white stone samples coated with calcium oxalate coating materials (Figure 6). The slight yellowing of the treated samples after 3 months is associated with dust deposition and the pollutant will be washed away by rain.

Figure 6. Field exposure results after one year.

Micro scale surface section of the field exposure sample was investigated with Raman spectral area map and SEM (Figure 7). Obvious evidence of the existence of the calcium oxalate was found; more specifically, the lightness of the red area correlates with the amount of calcium oxalate. After the field exposure, calcium oxalate is the dominant phase of sample surface, small quant of dark area often related to the surface roughness of sample. As shown in Figure 7c, calcium oxalate crystals of different sizes accumulate on the rough sample surface, and porosities are rarely present. The results show the effectiveness and durability of this treatment method for practical outdoor stone conservation.

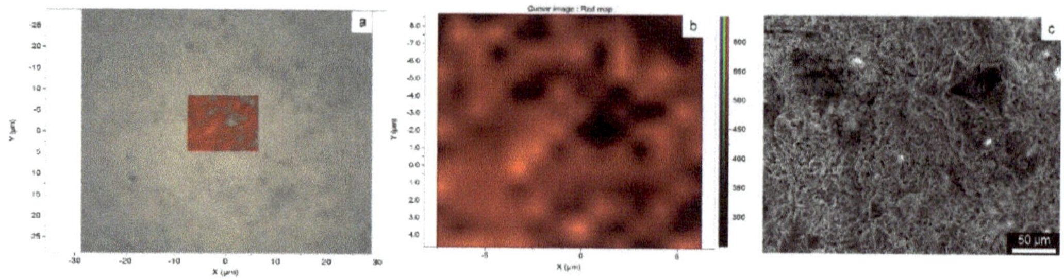

Figure 7. Field exposure results after one year. (**a**) Micrograph, (**b**) Raman mapping, (**c**) SEM.

4. Conclusions

In this study, the sequential application of ammonium oxalate, methyl oxalate neutral mixed, and calcium acetate acid mixed solutions was investigated as a possible route to provide carbonate rock with acid resistance properties and durable conservation at the same time.

The resulting coating exhibits a homogenous and refined microstructure, presumably thanks to the slowly hydrolysis of methyl oxalate neutral mixed and the effect of calcium acetate acid mixed crystal on crystal structure modification. After prolonged exposure to simulated acid rain (pH 2–4) and a selective outdoor environment, the acid resistance properties of the coated stone exhibit greater durability. This is possible due to the bonding force between the previous deposition particle and the sequential deposition layer. In this way, sequential deposition particles are firmly incorporated into the previous layer, thereby forming a durable conservation layer.

Considering the durability of the coating investigated in this study, the advantage of the sequential application is evident. More experiments to assess the mechanism of the coating formation process will be the next step in the research.

Supplementary Materials: The following supporting information can be downloaded at: https://www.mdpi.com/article/10.3390/coatings12101412/s1, Figure S1: The AFM of dolomite marble after ammonium oxalate solution treatment; Figure S2: SEM images of (a) Marble, (b) Sample 'SF', (c) Sample 'PF' after immersion tests for 6 h; Figure S3: SEM images after UT (a,d), AO (b,e) and 0.02 mol/L CA (c,f) were treated with 750 mL acid solution(HNO_3:H_2SO_4 = 1:1), with an initial pH of 4.0.

Author Contributions: Conceptualization, J.Z.; funding acquisition, S.W. and Q.M.; methodology, J.Z. and Y.G.; resources, S.W.; software, Y.G.; writing—original draft, J.Z.; writing—review and editing, H.H. and A.W. All authors have read and agreed to the published version of the manuscript.

Funding: This research was funded by Beijing Municipal Administration of Cultural Heritage (PXM2020-039208-000005).

Institutional Review Board Statement: Not applicable.

Informed Consent Statement: Not applicable.

Data Availability Statement: Not applicable.

Acknowledgments: The authors want to thanks Beijing Stone Carving Art Museum for the supply of material support in this work.

Conflicts of Interest: The authors declare no conflict of interest.

References

1. Liu, Y.; Dong, T.L.; Zhang, K.; Yang, F.W.; Wang, L.Q. Preliminary Study of the Targeted Cleaning of an Artificial Gypsum Layer on White Marble. *Coatings* **2021**, *11*, 37. [CrossRef]

2. Gherardi, F. Current and Future Trends in Protective Treatments for Stone Heritage. In *Conserving Stone Heritage: Traditional and Innovative Materials and Techniques*; Gherardi, F., Maravelaki, P.N., Eds.; Springer International Publishing: Cham, Switzerland, 2022; pp. 137–176.
3. Zucchelli, M.; Mazzon, G.; Bertolacci, L.; Carzino, R.; Zendri, E.; Athanassiou, A. Stone sustainable protection and preservation using a zein-based hydrophobic coating. *Prog. Org. Coat.* **2021**, *159*, 11. [CrossRef]
4. Ion, R.M.; Iancu, L.; Vasilievici, G.; Grigore, M.E.; Andrei, R.E.; Radu, G.I.; Grigorescu, R.M.; Teodorescu, S.; Bucurica, I.A.; Ion, M.L.; et al. Ion-Substituted Carbonated Hydroxyapatite Coatings for Model Stone Samples. *Coatings* **2019**, *9*, 231. [CrossRef]
5. Liu, Y.; Yang, F.W.; Zuo, G.F.; Zhang, R.X.; Wei, G.F.; Ma, Q. Protection of the surface weathering stone artworks by a chemical conversion method. *Constr. Build. Mater.* **2018**, *182*, 210–214. [CrossRef]
6. Dei, L.; Salvadori, B. Nanotechnology in cultural heritage conservation: Nanometric slaked lime saves architectonic and artistic surfaces from decay. *J. Cult. Herit.* **2006**, *7*, 110–115. [CrossRef]
7. Price, C.A. The consolidation of limestone using a lime poultice and limewater. *Stud. Conserv.* **1984**, *29*, 160–162. [CrossRef]
8. Drdacky, M.; Slizkova, Z.; Ziegenbalg, G. A Nano Approach to Consolidation of Degraded Historic Lime Mortars. *J. Nano Res.* **2009**, *8*, 13–22. [CrossRef]
9. Ziegenbalg, G.; Bra, K.; Pianski, J. Nano-lime-A new material for the consolidation and conservation of historic mortars. In Proceedings of the 2nd Historic Mortars Conference, Prague, Czech Republic, 22–24 September 2010; pp. 1301–1309.
10. Ziegenbalg, G. Colloidal calcium hydroxide-a new material for consolidation and conservation of carbonatic stones. In Proceedings of the 11th International Congress on Deterioration and Conservation of Stone III, Torun, Poland, 15–20 September 2008; pp. 1109–1115.
11. Rampazzi, L. Calcium oxalate films on works of art: A review. *J. Cult. Herit.* **2019**, *40*, 195–214. [CrossRef]
12. Sassoni, E. Hydroxyapatite and Other Calcium Phosphates for the Conservation of Cultural Heritage: A Review. *Materials* **2018**, *11*, 557. [CrossRef]
13. Lu, R.C.; Wang, L.; Liu, Y.; Yang, F.W.; Yang, L.; Wang, L.Q.; Gao, X. Conservation of surface gypsification stone relics using methanol solution of barium hydroxide as a novel treating agent. *Appl. Phys. A Mater. Sci. Process.* **2022**, *128*, 8. [CrossRef]
14. Fico, D.; Rizzo, D.; Casciaro, R.; Corcione, C.E. Historically Accurate Reconstruction of the Materials and Conservation Technologies Used on the Facades of the Artistic Buildings in Lecce (Apulia, Italy). *Materials* **2022**, *15*, 3658. [CrossRef] [PubMed]
15. Aragoni, M.C.; Giacopetti, L.; Arca, M.; Carcangiu, G.; Columbu, S.; Gimeno, D.; Isaia, F.; Lippolis, V.; Meloni, P.; Ezquerra, A.N.; et al. Ammonium monoethyloxalate (AmEtOx): A new agent for the conservation of carbonate stone substrates. *New J. Chem.* **2021**, *45*, 14. [CrossRef]
16. Dreyfuss, T. Artificially induced calcium oxalate on limestone in urban environments-New findings. *J. Cult. Herit.* **2020**, *42*, 56–63. [CrossRef]
17. Gomez-Villalba, L.S.; Feijoo, J.; Rabanal, M.E.; Fort, R. In-situ electrochemical synthesis of inorganic compounds for materials conservation: Assessment of their effects on the porous structure. *Ceram. Int.* **2021**, *47*, 30406–30424. [CrossRef]
18. Calore, N.; Botteon, A.; Colombo, C.; Comunian, A.; Possenti, E.; Realini, M.; Sali, D.; Conti, C. High Resolution ATR μ-FTIR to map the diffusion of conservation treatments applied to painted plasters. *Vib. Spectrosc.* **2018**, *98*, 105–110. [CrossRef]
19. Suyamud, B.; Ferrier, J.; Csetenyi, L.; Inthorn, D.; Gadd, G.M. Biotransformation of struvite by Aspergillus niger: Phosphate release and magnesium biomineralization as glushinskite. *Environ. Microbiol.* **2020**, *22*, 1588–1602. [CrossRef]
20. Maiore, L.; Aragoni, M.C.; Carcangiu, G.; Cocco, O.; Isaia, F.; Lippolis, V.; Meloni, P.; Murru, A.; Tuveri, E.; Arca, M. Synthesis, characterization and DFT-modeling of novel agents for the protection and restoration of historical calcareous stone substrates. *J. Colloid Interface Sci.* **2015**, *448*, 320–330. [CrossRef]
21. Ruiz-Agudo, E.; Alvarez-Lloret, P.; Putnis, C.V.; Rodriguez-Navarro, A.B.; Putnis, A. Influence of chemical and structural factors on the calcite-calcium oxalate transformation. *Crystengcomm* **2013**, *15*, 9968–9979. [CrossRef]
22. Burgos-Cara, A.; Putnis, C.V.; Ortega-Huertas, M.; Ruiz-Agudo, E. Influence of pH and citrate on the formation of oxalate layers on calcite revealed by in situ nanoscale imaging. *Crystengcomm* **2017**, *19*, 3420–3429. [CrossRef]
23. Conti, C.; Aliatis, I.; Casati, M.; Colombo, C.; Matteini, M.; Negrotti, R.; Realini, M.; Zerbi, G. Diethyl oxalate as a new potential conservation product for decayed carbonatic substrates. *J. Cult. Herit.* **2014**, *15*, 336–338. [CrossRef]
24. Xu, Y.; Chen, Q.; Zhao, Y.J.; Lv, J.; Li, Z.H.; Ma, X.B. Autocatalytic Kinetic Study of Dimethyl Oxalate Consecutive Hydrolysis. *Ind. Eng. Chem. Res.* **2014**, *53*, 4207–4214. [CrossRef]
25. Graziani, G.; Sassoni, E.; Franzoni, E.; Scherer, G.W. Hydroxyapatite coatings for marble protection: Optimization of calcite covering and acid resistance. *Appl. Surf. Sci.* **2016**, *368*, 241–257. [CrossRef]
26. Graziani, G.; Sassoni, E.; Scherer, G.W.; Franzoni, E. Resistance to simulated rain of hydroxyapatite- and calcium oxalate-based coatings for protection of marble against corrosion. *Corrosion Sci.* **2017**, *127*, 168–174. [CrossRef]
27. Maravelaki, P.N.; Kapetanaki, K.; Stefanakis, D. TEOS-PDMS-Calcium Oxalate Hydrophobic Nanocomposite for Protection and Stone Consolidation. *Heritage* **2021**, *4*, 4068–4075. [CrossRef]
28. He, L.; Wang, L.; Liu, Y.; Yang, F.W.; Gao, X. Preparation of calcium oxalate film on the surface of historic carbonate stone in a humid environment. *New J. Chem.* **2022**, *46*, 1099–1104. [CrossRef]
29. Mendes, G.D.; Bahri-Esfahani, J.; Csetenyi, L.; Hillier, S.; George, T.S.; Gadd, G.M. Chemical and Physical Mechanisms of Fungal Bioweathering of Rock Phosphate. *Geomicrobiol. J.* **2021**, *38*, 384–394. [CrossRef]

30. Aggelakopoulou, E.; Bakolas, A. What were the colors of the Parthenon? Investigation of the entablature's cornice blocks. *J. Archaeol. Sci.* **2022**, *140*, 18. [CrossRef]
31. Akbar, N.A.; Aziz, H.A.; Adlan, M.N. Characteristic Study of Physical and Chemical Properties of Limestone Subjected to Adsorption of Fe and Mn in Groundwater. In Proceedings of the AWAM International Conference on Civil Engineering (AICCE), Nibong Tebal, Malaysia, 21–22 August 2019; pp. 621–627.
32. Ibis, F.; Dhand, P.; Suleymanli, S.; van der Heijden, A.; Kramer, H.J.M.; Eral, H.B. A Combined Experimental and Modelling Study on Solubility of Calcium Oxalate Monohydrate at Physiologically Relevant pH and Temperatures. *Crystals* **2020**, *10*, 924. [CrossRef]

coatings

MDPI

Article

A Novel Protection Method for Carbonate Stone Artifacts with Gypsum Weathering Crusts

Ruicong Lu [1,2,3], Lu He [1,2,3], Ting Li [1,2,3], Fuwei Yang [1,2,3,*], Yan Liu [1,2,3], Kun Zhang [1,2,3] and Xinnan Chen [1,2,3]

1 China-Central Asia "the Belt and Road" Joint Laboratory on Human and Environment Research, Northwest University, Xi'an 710127, China
2 Key Laboratory of Cultural Heritage Research and Conservation, Northwest University, Xi'an 710127, China
3 School of Culture Heritage, Northwest University, Xi'an 710127, China
* Correspondence: yangfuwei@nwu.edu.cn

Abstract: An innovative method using a methanol solution of barium hydroxide-urea as a protective agent was investigated for the conservation of stone artifacts with harmful gypsum weathering crusts. In this method, the methanol solution of barium hydroxide-urea and water were introduced into the gypsum crust in sequence by surface spraying. By doing so, the harmful gypsum crust is directly converted into a barium sulfate—calcium carbonate composite protective layer. The properties of the composite layer were characterized by SEM-EDX, XRD, ATR-FTIR, IC, water solubility, wetting angle, color difference, open porosity, capillary water absorption, and surface hardness. The results of the morphological and composition characterization (SEM-EDX, XRD, ATR-FTIR) indicate that the added urea can promote the carbonization reaction effectively. In addition, the methanol solution of barium hydroxide-urea can penetrate deep into the gypsum crust. The results of the physical properties characterization denote that the water stability of the specimens was significantly increased after the protection treatment; an approximate ten-fold stronger water resistance ability was achieved. Meanwhile, the intrinsic physical properties of gypsum crust, such as pore structure and original appearance, could basically be maintained. The presented conservative method has high facility and controllability and satisfying conservation effect, which means it has potential in the conservation of surface weathering carbonate stone artifacts.

Keywords: barium hydroxide; urea; methanol solution; gypsum crust; stone artifacts

Citation: Lu, R.; He, L.; Li, T.; Yang, F.; Liu, Y.; Zhang, K.; Chen, X. A Novel Protection Method for Carbonate Stone Artifacts with Gypsum Weathering Crusts. *Coatings* **2022**, *12*, 1793. https://doi.org/10.3390/coatings12111793

Academic Editor: Enrico Quagliarini

Received: 10 October 2022
Accepted: 18 November 2022
Published: 21 November 2022

1. Introduction

Calcareous stone artifacts are widespread around the world. However, the stability of such artifacts is not strong enough [1,2]. As a result, they are susceptible to corrosion, and weathering diseases are prevalent among them. There are a number of factors that can cause this damage, and the effects of sulfur oxides are non-negligible. With the development of industrialization, the contents of these kinds of chemicals have increased dramatically and will react with the calcareous stone artifacts [3]. This will transform calcium carbonate into calcium sulfate, which will eventually generate the gypsum crust [4]. The resulting gypsum crust is believed to be one of the most destructive weathering products for contaminated artifacts. This mainly results from the much higher solubility of gypsum in water ($Ksp = 3.14 \times 10^{-5}$) than that of calcium carbonate ($Ksp = 3.36 \times 10^{-9}$) [5]. Thus, the surface erosion of stone artifacts with gypsum weathering crusts can be readily caused by rain scouring. In addition, the higher solubility of gypsum can lead to serious flaking and crumbling damage of the stone artifacts because of the re-crystallization pressure [6,7].

Different types of methods have been studied to conserve the stone artifacts, and several new approaches have been proposed in recent years. Removing the gypsum crust directly via cleaning is a traditional method and various techniques have been investigated by scientists, such as laser cleaning, chemical cleaning, and bio-cleaning [8–10]. Recently,

Yan Liu et al. (2021) put forward a novel method using the scavenger of barium carbonate embedded in an absorbent cotton-coating; they let it react with the calcium sulfate specifically to achieve a targeted cleaning outcome on a surface sulfated marble [11]. Cleaning is an effective method. However, in most cases, the surfaces of the stone artifacts always contain different decorative elements or characteristics that carry important historical and cultural data [12]. Therefore, since the losses of historical and cultural values are irreversible, the gypsum crust cannot be simply removed in its entirety. With the development of organic materials, they have been applied in many areas, including the conservation of deteriorated stone artifacts. Commercial acrylic and methacrylic resins have been extensively applied since the 1960s [13]. However, these conventional organic polymers are incompatible with the inorganic substrates protected and show distinct aging phenomena, such as yellowing, cracking, and embrittlement after a period of time [14]. In recent years, inorganic components have been added to improve the properties of pure organic polymers. Many different kinds of inorganic nanometric reinforcing agents, such as TiO_2 and SiO_2, were incorporated with organic polymers by scientists who sought better conservation material [15]. Nevertheless, recent results showed that the added TiO_2 nanoparticles can cause the degradation of the incorporated Paraloid B-72 after exposure to UV light for a certain period of time; other outcomes also indicate that the durability of these organic–inorganic hybrid materials are not satisfying [16]. Conversely, the inorganic materials show high compatibility with the protected inorganic substrates; thus, leading to high durability. Lime water and barium hydroxide aqueous solution were the earliest-studied inorganic consolidants [17]. Nevertheless, their conservation effects are also dissatisfactory. The use of lime water was previously estimated to be inefficient, mainly due to the low solubility of calcium hydroxide in the water. Even with a massive amount of time of the conservation treatment, the obtained conservation effect is still very poor [18]. Theoretically, the barium hydroxide aqueous solution can convert the hazardous gypsum crust into barium sulfate and calcium hydroxide, which could gradually be carbonized to calcium carbonate. These acquired substances will provide protective effects against the decay of the stone [17]. Nonetheless, the extremely high reactivity of barium hydroxide in an aqueous solution makes this material unsatisfactory. When it contacts carbon dioxide in the atmosphere and the gypsum (or another sulfate, such as sodium sulfate in the interior of the stone artifacts), barium carbonate, barium sulfate, calcium hydroxide, as well as other chemicals would yield rapidly [19]. These substances will hinder the barium hydroxide aqueous solution from deeper penetration into the substrate, leading to a relatively restricted conservative effect presented only on the very surface of the substrate rather than providing sustained effective protection [20]. Hence, the Ferroni–Dini method was developed to obtain better conservation results. This method consists of two sequential procedures: desulfuration by a solution of ammonium carbonate and consolidation by an aqueous solution of barium hydroxide. The added desulfuration procedure aims to reduce the amount of the sulfate on the surface of the infected cultural heritage to ensure the better ability of the barium hydroxide aqueous solution to penetrate deep into the substrate and obtain a better conservation effect [21]. However, after this treatment, different outcomes occur on different kinds of cultural heritages. It achieves positive results on the frescoes and, inversely, negative consequences on the stone artifacts [22,23]. The reason for this should be that the gypsum crust on frescoes is thin and could obtain complete desulfuration [24]. On the contrary, the gypsum crust on stone artifacts is too thick to obtain complete desulfuration [25]. Therefore, the undesired reactions mentioned above will inevitably occur on stone artifacts with gypsum weathering crusts and lead to unsatisfactory protective properties. Thus, some new kinds of inorganic materials have also recently been explored, such as hydroxyapatite, nano calcium hydroxide, and barium hydroxide dispersions [26,27]. Nevertheless, the shortcomings of these materials are also evident. The introduction of hydroxyapatite would provide an abundant source of phosphorus in favor of many microorganisms. As for the nano organic materials, the preparation methods are mostly complicated [28]. In addition, nanoparticles tend to agglomerate; this will constrain their penetrability. Hence, the conservation effect can be impaired, too.

With all the limitations of the materials mentioned above, recently, we proposed the methanol solution of barium hydroxide as a novel treatment agent for the conservation of stone artifacts with gypsum weathering crust; the results were satisfying because of the organic solvent employed. When barium hydroxide is dissolved in a methanol solvent, it does not possess the activity to react with carbon dioxide in the atmosphere and calcium sulfate in the gypsum crust, as they are ionic reactions that can only occur in aqueous solutions. As a result, insoluble substances, i.e., barium carbonate, barium sulfate, and calcium carbonate, are not produced, ensuring that the solution has the ability to penetrate deep into the substrate. Moreover, the solubility of the barium hydroxide in the methanol solvent is much higher and the preparation method of the new treating solution is fairly facilitated [29]. However, this approach may face obstacles when applied to outdoor calcareous stone artifacts due to difficulties in further carbonating the resulting calcium hydroxide. Because the methanol solution of barium hydroxide can delve deep into the substrate, the carbonization reaction of the calcium hydroxide produced in the inner part of the substrate could be restricted since the content of the carbon dioxide in the atmosphere is quite low (0.038%) and the surface reaction will take place first. Thus, the complete carbonation reaction of outdoor stone artifacts after treatment with a barium hydroxide methanol solution may take a long time. Hence, in this paper, we prepare a methanol solution of barium hydroxide-urea as a new reagent for the conservation of stone artifacts with gypsum weathering crusts. The addition of urea could provide a continuous source of carbonate ions and permit a continuous carbonation reaction without compromising the advantages of the barium hydroxide methanol solution. Therefore, the ability of the methanol solution of barium hydroxide-urea to penetrate deep into the gypsum crust, as well as the high controllability (the reaction only occurs after the water supply) of this application strategy, will be maintained; they are of great importance for the conservation of stone artifacts with gypsum weathering crusts. In addition, due to the addition of urea, the entire conservation reaction can take place from the interior of the substrate without restriction, and the unstable intermediate product calcium hydroxide will be eliminated as soon as possible. Hence, a good conservation effect will be ensured. In application, the methanol solution of barium hydroxide-urea was first applied into the gypsum crust at different times to compare the conservation effects of the different amounts of introduced protectants. Water was then provided to stimulate reactions between them. The simple application procedures of the proposed protectants are shown in Figure 1. The conservation effect was evaluated by SEM-EDX, XRD, ATR-FTIR, water solubility, color difference, IC, wetting angle, open porosity, capillary water absorption, and surface hardness.

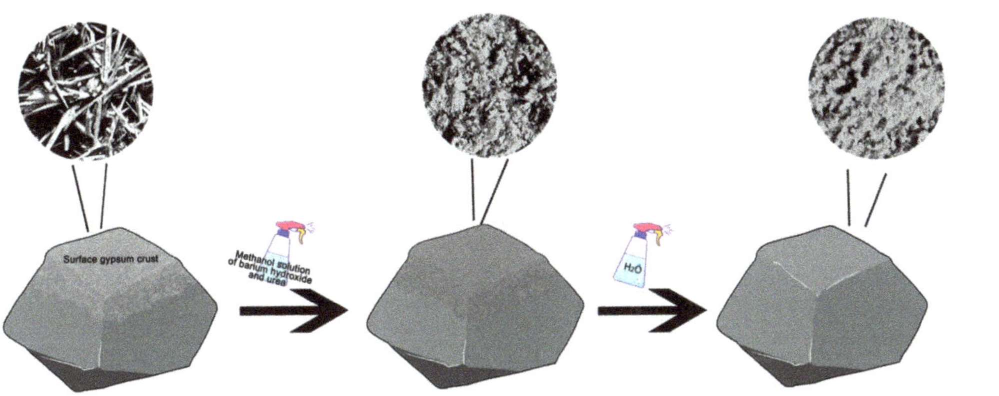

Figure 1. Conservation schematic diagram of surface gypsification stone artifacts.

2. Materials and Methods

2.1. Sample Preparation

All chemicals were of analytical grade and used without further purification. $Ba(OH)_2$, urea, and anhydrous methanol were purchased from Sinopharm Group Co., Ltd., Shanghai, China.

Referring to the previous literature, the specimens were prepared by the following method: the marble specimens (2.0 × 2.0 × 2.0 cm) were cleaned with deionized water first and then left to dry for one day. Afterward, to simulate a surface gypsum crust, the specimens were placed into an aqueous solution of H_2SO_4 at pH 2 for 24 h [26]. Then, the specimens were taken out and further desiccated prior to use.

The mixed solution of barium hydroxide-urea in methanol at a concentration of 10% was prepared by dissolving barium hydroxide and urea into the anhydrous methanol. After stirring for hours, the solution was centrifuged and ready for use.

The conservation routine was as follows: the specimens were sprayed with the reagent solution until they could no longer be absorbed and then left to dry in a confined environment. In order to compare the conservation effects of the different amounts of introduced protectants, the above operation was operated three successive times on three sets of specimens, each containing five specimens. Later, the specimens were provided with sufficient water by spraying, and further cured for one month in a sealed and moist environment before various tests were carried out. The protective performances of the specimens with different numbers of treatments were characterized by water solubility, color difference, IC, wetting angle, open porosity, capillary water absorption, and surface hardness. To further characterize the conversion process of the conservation treatment, SEM, XRD, and FTIR were performed on 3-fold-treated specimens, from which the best conservation effects were obtained.

2.2. Characterization

The structure and composition of the specimens were analyzed by scanning electron microscopy (SEM, VEGA-3XMU, TESCAN, Brno, Czech, BSE mode, 12.00 kV of accelerating voltage, and 10.0 mm of working distance) with energy-dispersive X-ray spectroscopy (EDX, Genesis 2000 XMS, EDAX, Inc., Mahwah, NJ, USA), X-ray diffractometry (XRD, Smart Lab, Rigaku, Tokyo, Japan, Cu Ka radiation, scanning range 10°~90°, step length 0.01°, scanning speed 10°/min), and attenuated total reflectance–Fourier transform infrared (ATR-FTIR, TENSOR 27, Bruker, Billerica, MA, USA, scanning wave number 4000–600 cm^{-1}, resolution 4 cm^{-1}, scanning times 16 times) spectroscopy.

The water solubility of the specimens with different times of the conservation treatment was approximately measured by a conductivity meter (DDSJ-308F, Rex, INESA Scientific Instrument Co., Ltd., Shanghai, China). The specimens were placed in a beaker containing 200 mL of water under mechanical agitation, and the conductivity values of the solution were recorded automatically every minutes as soon as the specimens were placed in the beaker until the values were approximately steady. Ion chromatography (IC, ICS 1100/1600, Thermo Fisher Scientific, Cleveland, OH, USA) was applied to further affirm the ion content of the solution after the detection of the water solubility of the specimens.

To assesses the water repellency of the specimens, the wetting angles of the specimens were tested via a contact angle tester (JGW-360B, Beijing Haifuda Technology Co., Ltd., Beijing, China).

To evaluate the color differences of the specimens, a portable colorimeter (WSC-2B, INESA Scientific Instrument Co., Ltd., Shanghai, China) was used. The treated specimens after curation were further desiccated under dry and ventilated conditions for 3 days and were then tested. Five points from the different parts of the surfaces of each blank and treated specimen were tested to minimize the experimental error. The results were analyzed according to the colorimetric coordinates (CIELAB); the formula is as follows:

$$\Delta E^* = (\Delta L^{*2} + \Delta a^{*2} + \Delta b^{*2})^{1/2} \quad (1)$$

According to this color representation, L* is lightness, a* is the red–green component, and b* is the yellow–blue component [30].

To determine the capillary water absorption and open porosity of the specimens, an electronic density tester (MZ-C300, Mayzun, Shenzhen, China) was employed.

To invest in the surface hardness of the specimens, a shore D durometer (LX-D-1, Dongguan SanLiang measuring tools Co., Ltd., Dongguan, China) was applied.

3. Results and Discussion

3.1. Morphological and Chemical Characterization

The surface microstructure changes of the specimens are shown in Figure 2. As expected, a loosened needle-like morphology appeared for the blank specimen (Figure 2a,d). The micropores are clearly visible, as depicted in Figure 2d. This typical gypsum morphology has been confirmed by many scientists before [31,32]. The characteristic diffraction peaks (d = 7.6057, 4.2649, 3.7912, 3.0515, 2.8643, 2.6760 Å) in the XRD results (Figure 3a) further corroborated the gypsum component. After introducing only the methanol solution of barium hydroxide-urea, it can be seen in Figure 2e that the labeled in-compact protectants scattered across the entire surface act similarly to padding rather than provide any connection strength. The voids and pores of the original blank specimens were completely covered by them (Figure 2b,e). For the XRD results, gypsum is still the only phase that could be observed and the existence of the barium-contained phases could not be detected (Figure 3b). The reason for this could be that the barium-containing phases were in amorphous form and the XRD representation was not able to identify these types of entities. This denotes that the reaction between the reagent solution and the gypsum crust will not occur without the water supply. This is because the reactions between them are ionic and cannot occur in organic solvents. After the supply of water, however, there was an evident change in the morphology of the treated specimens. The in-compact padding substance before the water supply changed into a more continuous structure. The scattered introduced conservative materials and the original gypsum matrix merged entirely (Figure 2c,f). The XRD results further reveal that the compositions of the newly emerged structure are barium sulfate (d = 3.8898, 3.4333, 3.3106, 2.8376, 2.1008 Å) and calcium carbonate (d = 3.0362, 2.2847, 1.9111, 1.8748 Å) (Figure 3c). Thus, the new compact structure presented in Figure 2c,f is a barium sulfate-calcium carbonate composite layer. In addition, the crystallographic form of the calcium carbonate is calcite; this follows the investigation results of the previous studies, i.e., the addition of water supply is more likely to produce calcite. On the contrary, when there is a large number of organic molecules in the solution and no straight water supply, this shows a tendency to favor the formation and kinetic stabilization of vaterite instead of stable calcite [33,34]. These results indicate that the conservation reaction between barium hydroxide and the gypsum crust is completed after the water supply. Because of the introduction of water, the urea was decomposed into the form of ammonia and carbon dioxide. The ammonia was released into the environment; the carbon dioxide produced via this process was performed as a continuous source of carbonate ions to lead to the carbonation reaction from the internal substrate. The chemical equations are as follows:

$$CO\,(NH_2)_2 + H_2O = CO_2 + 2NH_3 \tag{2}$$

$$Ba(OH)_2 + CaSO_4 = BaSO_4 + Ca(OH)_2, \tag{3}$$

$$Ca(OH)_2 + CO_2 = CaCO_3 + H_2O \tag{4}$$

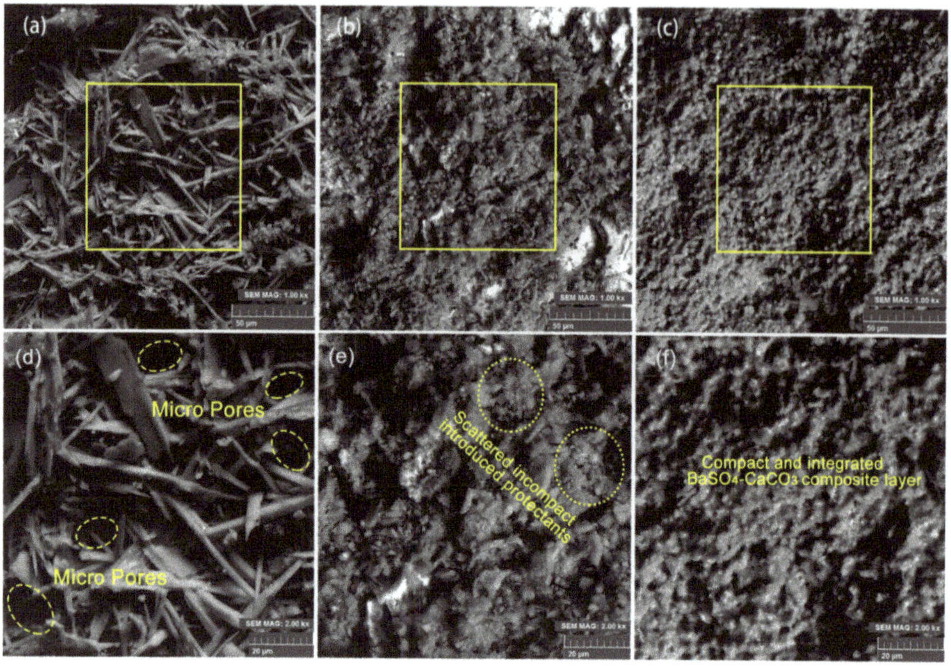

Figure 2. Surface morphology of the specimens with different magnifications (the yellow rectangle areas of the (**a**–**c**) are magnified to (**d**–**f**), respectively), (**a**,**d**): blank specimen, (**b**,**e**): treated specimen before water supply, (**c**,**f**): treated specimen after water supply.

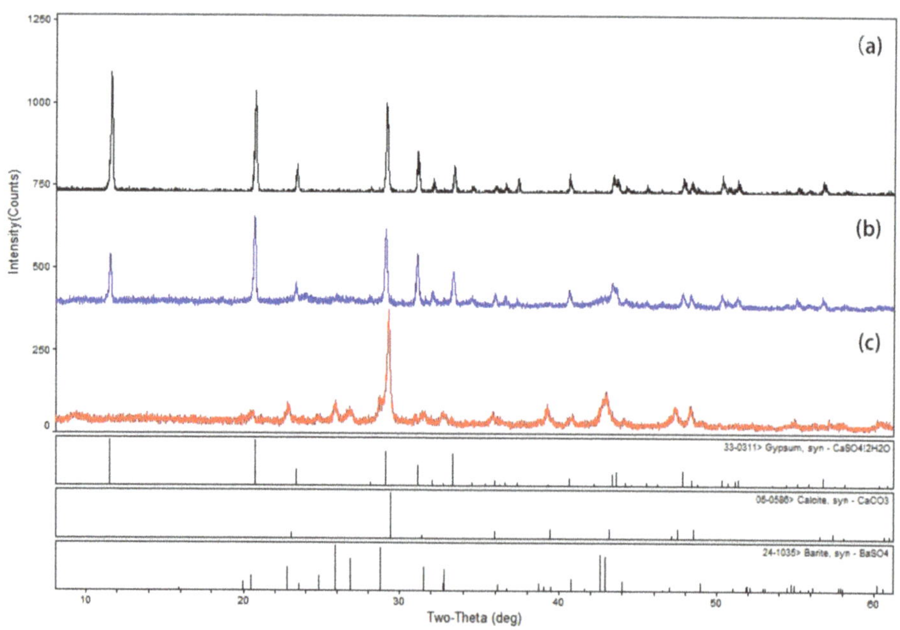

Figure 3. XRD results of the specimens: (**a**) blank specimen, (**b**) treated specimen before water supply, (**c**) treated specimen after water supply.

The FTIR spectra of the specimen after treatment with the reagent solution and before the water supply are presented in Figure 4. Moreover, the spectra of the gypsum, barium hydroxide, and urea are displayed for comparison. As shown in Figure 4, the characteristic peaks of the gypsum matrix can be clearly observed from the spectrum of the specimen. The general trend of peaks around 2250 cm^{-1} and 2800 cm^{-1} in the spectrum of the specimen is consistent with the spectrum of barium hydroxide, suggesting the presence of barium hydroxide. Meanwhile, the absence of the sharp vibrational band of OH stretching modes in $Ba(OH)_2$ located at 3570 cm^{-1} in the specimen's spectrum should be attributed to the reaction between barium hydroxide and methanol [35,36]. The peaks that appeared at around 1620 and 1460 cm^{-1} can be assigned to amide II bending and C-N stretching vibrations of the urea [37]. In addition to the peaks mentioned above, other peaks are marked in the figure and can further confirm the successful introduction of barium hydroxide and urea into the specimen, which cannot be verified in the XRD results. The FTIR results for the specimen after the water supply are depicted in Figure 5. Similar to Figure 4, the spectra of the potential (definite) components of the specimen are also presented. Even though the spectra of the calcite and barium carbonate are highly similar, the sharp peaks located at 1797, 874, and 711 cm^{-1} in the spectrum of the specimen can easily be attributed to calcite, as their specific locations are quite different from those of the barium carbonate peaks. However, the peak for mode v_3 of the CO_3^{2-} is located at 1454 cm^{-1} in the specimen's spectrum, which is slightly different from the spectrum of the calcite. Nevertheless, the slight discrepancy of this peak is not crucial for the confirmation of the calcite as it can shift occasionally; this can also be observed in the other forms of calcium carbonate. Conversely, the peaks of the other forms of the calcium carbonate located at lower wavenumbers, such as calcite (874 and 711 cm^{-1}), aragonite (856 and 711 cm^{-1}), and vaterite (875 and 745 cm^{-1}), are usually not changed under different situations [33,38–40]. Hence, the existence of the calcite can be fully ascertained here; this is in accordance with the XRD results. The appearance of the peaks at 637 and 610 cm^{-1} in the spectrum of the specimen, when compared with that of the gypsum, should certainly be ascribed to barium sulfate. In addition, the absorption peaks at 1080 cm^{-1} in the specimen's spectrum come from the sulfur-oxygen (S-O) stretching of the barium sulfate [41]. These FTIR analytical data are in good agreement with the XRD results and can further complete the results. In the XRD results, the urea and barium hydroxide cannot be distinguished from the spectrum before the water supply. Nevertheless, these substances can be clearly identified from the FTIR results. Therefore, the whole reaction process is fully confirmed.

To determine the length of time required for the conservation reaction to occur entirely, the FTIR was applied to identify the extent of the reaction since the reagent solution was introduced and the spectra are shown in Figure 6. For FTIR spectra, the quantitative analysis of different substances in the tested specimen was normally performed by comparing peak intensities. Hence, herein, the peak intensities of the characteristic peaks of the produced calcium carbonate (1454, 874 cm^{-1}) and barium sulfate (1080 cm^{-1}) can be used to ascertain the contents of them in the treated specimen and estimate the extent of the conservation reaction. By comparing the peak intensities of the calcium carbonate (874 cm^{-1}) and original gypsum crust (668 cm^{-1}), it is evident in Figure 6 that the peak intensity of calcium carbonate shows an incremental trend along with the time; however, it started barely to change after 4 days of the water supply. Meanwhile, the peak intensity of the barium sulfate (1080 cm^{-1}) also presented the same accumulative trending as that of calcium carbonate, which can be observed in Figure 6. In addition, the intensity of the peak located at 1454 cm^{-1} for the calcium carbonate almost did not alter at different times; this was observed in the preceding studies [40,42]. However, as depicted in Figure 6, the lengths of the vertical red lines may also stand for the content of the calcium carbonate. This is because the spectrum of the pure calcium carbonate shows a distinct downward feature at around 1400 cm^{-1} (Figure 5). The lengths of the red lines indicate the extent of this downward feature, which is the content of the calcium carbonate. It also increases during the first four days after the water supply and is almost unaltered thereafter. Therefore, the whole conservation reaction can take place within one week.

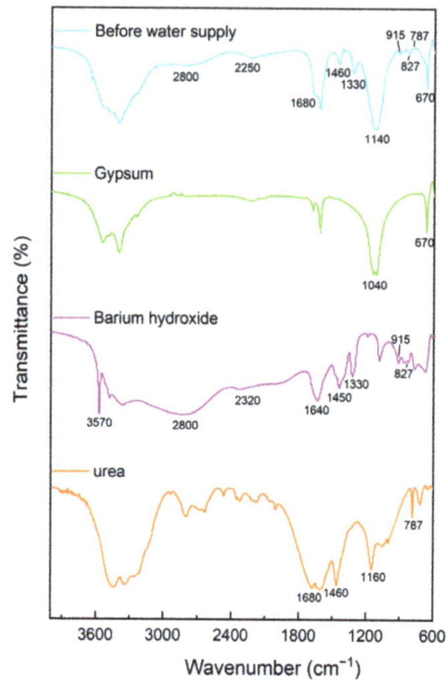

Figure 4. FTIR spectra of the specimen before the water supply.

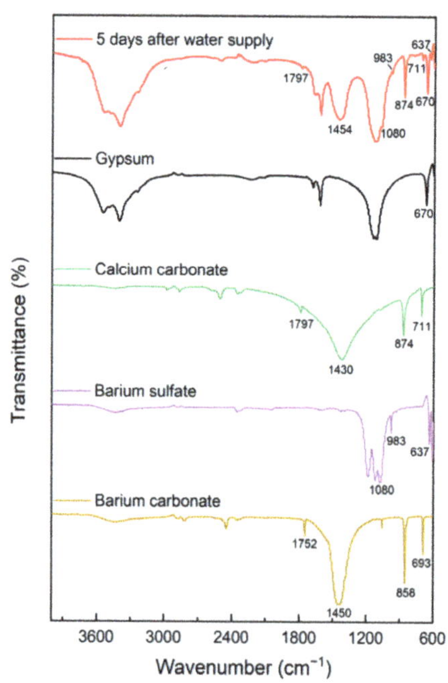

Figure 5. FTIR spectra of the specimen after the water supply.

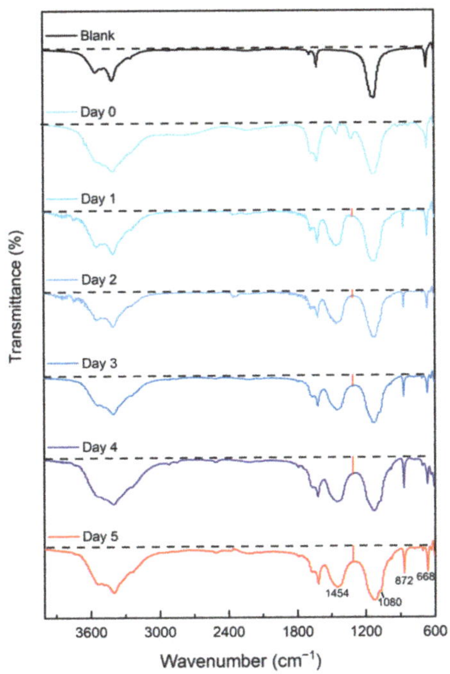

Figure 6. FTIR spectra of the specimens at different time intervals.

The cross-sectional morphology changes of the specimens are shown in Figure 7. The blank specimen displays a structure in agreement with the surface morphology (Figure 7a,d). After being treated with a reagent solution, and before being supplied with water, the morphology of the specimen showed a difference. As labeled in Figure 7e, the areas represented by the five yellow lines are a little darker than their surrounding areas and there are other different parts similar to them. In addition, the surrounding areas in lighter colors show cobweb-like morphology. It is clear that these features are not presented in the blank specimen and it should be the result from the introduced protectants. The surrounding areas in lighter colors should be the introduced protectants that cover the original gypsum substrate. However, the specimens need to be broken to observe the cross-sectional morphology. Through this process, the original morphology of the specimens can be affected. The gypsum substrate, which is supposed to be covered evenly, is partially exposed, and the exposed gypsum substrates can be observed as the darker areas labeled in Figure 7e. After the water was provided, a more compact structure emerged, and the gypsum substrate was hardly exposed (Figure 7c,f). The needle-like structure of the blank specimen is tightly folded by the newly produced material, which is barium sulfate and calcium carbonate according to the aforementioned XRD and FTIR results. The EDX results of the treated specimens before and after the water supply indicate that the reagent solution has the ability to penetrate deep into the substrate. The even distribution of the barium element can be detected consistently from the surface to a depth of about 3000 μm for the specimen provided with water (Figure 8right), which is a little deeper than the specimen before the water supply (Figure 8left). This may be due to some protective materials being transferred with the flowing water. These results imply that the ability of the methanol solution of barium hydroxide to penetrate deep into the substrate was not affected by the adding urea and it is still much better than the aqueous solution because of the organic solvent adopted [29].

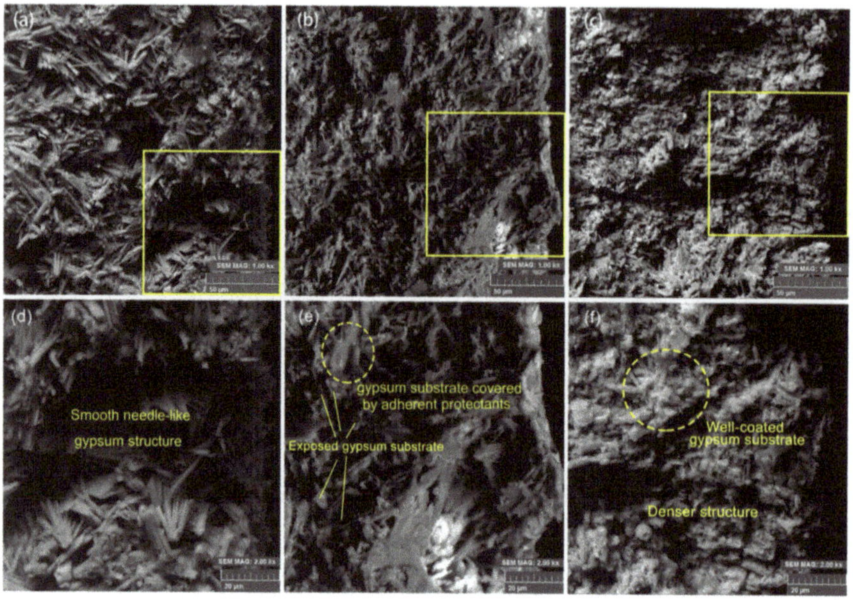

Figure 7. Cross-section morphology of the specimens with different magnifications (the yellow rectangle areas of the (**a–c**) are magnified to (**d–f**), respectively), (**a**,**d**): blank specimen, (**b**,**e**): treated specimen before water supply, (**c**,**f**): treated specimen after water supply.

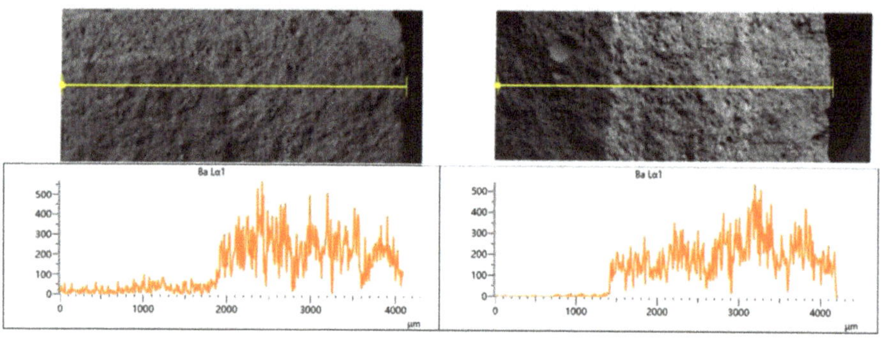

Figure 8. Line scan results of specimens; (**left**): the treated specimen before water supply, ((**upper**): SEM image with the yellow line analyzed by EDX, (**lower**): line scan results of the barium element), (**right**): treated specimen after water supply, ((**upper**): SEM image with the yellow line analyzed by EDX, (**lower**): line scan results of the barium element).

3.2. Physical Properties Characterization

Electrical conductivity was used to evaluate the water solubility of the specimens; the results are illustrated in Figure 9. It is commonly used for indicating the total concentration of the ionized constituents of a solution and is positively related to the sum of the anions (or cations) chemically determined [43]. Gypsum is a fairly dissoluble salt. Hence, when the blank specimen was placed into the water, an appreciable quantity of the gypsum was decomposed into calcium ions and sulfate ions, which made the electrical conductivity of the blank specimen reach about 1800 μS/cm. After the conservation operation, the less soluble barium sulfate and calcium carbonate will be produced and the original gypsum substrate will be covered, according to the SEM images. Thus, the conductivity of the treated specimens is significantly reduced. The maximum value of the conductivity of the specimen with only one spray operation is about 800 μS/cm and it could be further decreased to about 500 μS/cm after two operations. After triple-spray operations, the value is only about 250 μS/cm, which is far below the results of the blank specimen. The most likely causes for these different degressive results are that the barium hydroxide absorbed by the specimens (with different spray-operation times) is incremental and the gypsum substrate could be coated better.

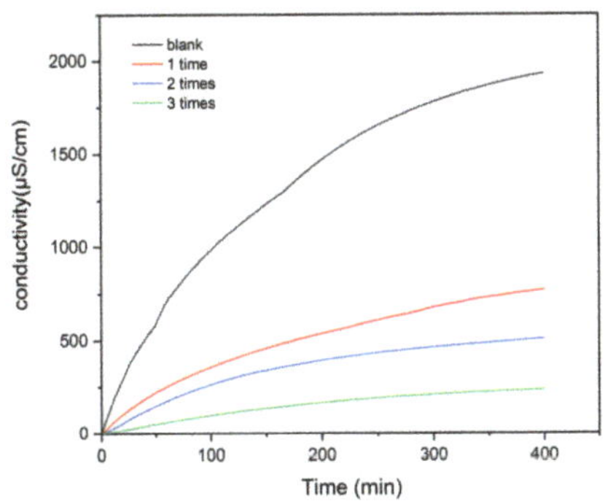

Figure 9. Conductivity results of the specimens before and after treatment.

The IC results are delineated in Figure 10 and the specific data values are exhibited in Table 1. Due to instrumental limitations, SO_4^{2-} is the only ionic species that can be efficiently detected and is relevant to this study. In general, the regularity of the variation of the SO_4^{2-} content among the different solutions is roughly consistent with the conductivity results. The solution of the blank specimen reaches its highest value and the solution of the specimen after three treatments achieves its lowest, as expected. As mentioned above, due to the difference in solubility between the original substrate gypsum and the product barium sulfate, the SO_4^{2-} in the solution should mainly originate from the gypsum. In addition, the SO_4^{2-} content ratio between the blank specimen and the specimen after 3 treatments is as high as 30 (Table 1). This is a larger number compared with the approximate 8 times calculated from the electrical conductivity results. The reason for this could be that there may be some contamination ions in the specimens or the urea was not decomposed completely. Furthermore, the produced calcium carbonate may also have led to this outcome. This significant reduction in SO_4^{2-} content indicates that the initially susceptible gypsum crust was almost fully shielded after three treatments with the methanol solution of barium hydroxide and urea.

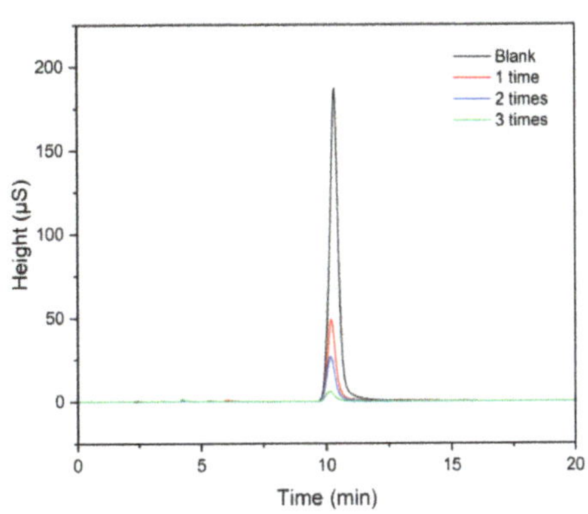

Figure 10. Ion chromatography results of the specimens before and after treatment.

Table 1. The SO_4^{2-} ion content of the specimens with different treatment numbers.

Specimens with Different Treatment Numbers	Type	Area (S*min)	Height (S)	Amount (mg/L)
0 (blank)	BMB	75.083	186.735	65.413
1	BMB	19.226	48.921	16.783
2	BMB	10.940	27.031	9.569
3	BMB	2.575	5.929	2.287

The outcomes of the wetting angles are displayed in Figure 11. All images were taken as soon as the droplets contacted the surface due to the high hydrophobicity of the gypsum substrate, after which, they were absorbed by the substrate. The wetting angle of the blank specimen was only 17.00° and increased slightly to 18.68° after being treated only once with the methanol solution of barium hydroxide and urea. After being treated twice, the wetting angle of the specimen changed to 30.00° and increased to 42.56° after three treatments. Typically, the magnitude of the wetting angle represents the hydrophilic or hydrophobic nature of the material. When the wetting angle of the surface is located in the range of 0° to 90°, the surface can be regarded as hydrophilic. Hence, the conservation treatment does not alter the hydrophilic nature of the original gypsum crust and the "breathing function" of the matrix could be maintained. The results show distinct differences with the organic coatings, which lead to a complete water repellency and are incompatible with the inorganic gypsum. Hence, this will eventually cause fissures and cracks [44,45]. However, despite the definition mentioned above, a larger wetting angle also represents a more hydrophobic surface even if it is an acute angle. Therefore, the surface of the specimen will become more hydrophobic along with the incremental number of conservation treatments. This could mainly be due to the generated barium sulfate and calcium carbonate. They have fairly low solubility and are able to cling tightly to the gypsum crusts, making them more durable and allowing for better water resistance.

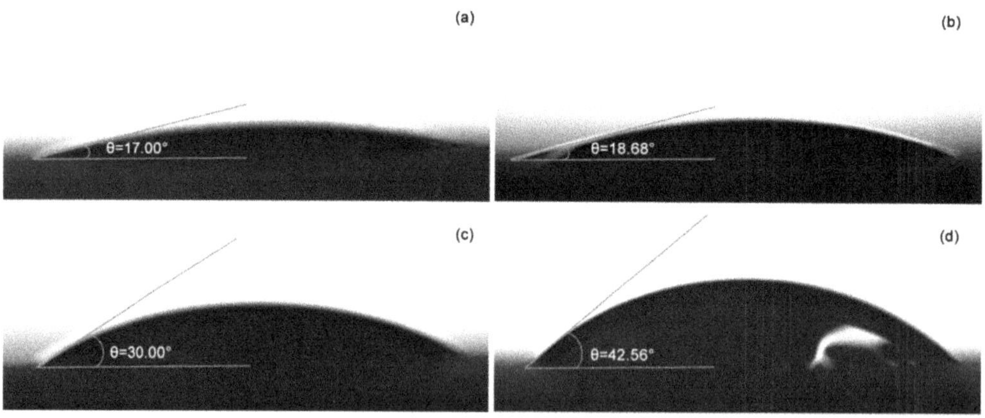

Figure 11. Wetting angles of the specimens with different treatment numbers, (**a**) blank specimen, (**b**) specimen after a single treatment, (**c**) specimen after two treatments, (**d**) specimen after three treatments.

The outcomes of the color differences of the specimens before and after treatment are shown in Table 2. According to the formula mentioned above, the color differences (ΔE*) of the treated specimens compared with the blank specimens were calculated. The results of the specimens with different treatment numbers are all varied in a range far below 5; this will not be recognized by the human eye [46]. It signifies that the "original appearance" of the specimen was almost unaffected after the introduction of the methanol solution of barium hydroxide-urea; this is of vital importance for cultural heritage objects.

Table 2. Color differences of the specimens before and after treatment.

Specimens with Different Treatment Numbers	L*	a*	b*	ΔL*	Δa*	Δb*	ΔE*
0 (blank)	38.98	0.99	3.90	-	-	-	-
1	39.14	0.88	3.34	0.16	−0.11	−0.56	0.59
2	38.65	0.92	3.28	−0.33	−0.07	−0.62	0.71
3	39.18	0.78	2.82	0.20	−0.21	−1.08	1.12

The results of the open porosity and capillary water absorption of the specimens are presented in Figures 12 and 13. The open porosity of the specimens decreased from about 40% to 34% after three treatments (Figure 12). This suggests that the specimens acquired denser and more compact structures after treatment, which is clearly shown in the SEM results. This could be mainly due to the adequate introduction of the methanol solution of barium hydroxide-urea. As a consequence of the formation of the inert protective barium calcium sulfate carbonate layer, the capillary water absorptions of the specimens reduced from about 30% to 23% (Figure 13). The extent of this variation is an evident change indeed. However, it is not unacceptable when compared with synthetic polymer film [47]. This is roughly consistent with the results obtained for the wetting angle: the hydrophilic nature of the primordial gypsum crust can be considered largely maintained, especially because the solubility of the specimens is significantly reduced.

The surface hardness of the specimens before and after treatment are shown in Figure 14. The average surface hardness of the blank specimens is about 69.1 HD, which increases to about 72.5 HD after three treatments. This is in accordance with the results of the SEM and the open porosity. After the introduction of the methanol solution of barium hydroxide-urea, the initial loose gypsum structure was wrapped rigidly by the producing barium sulfate and calcium carbonate. As a result, a denser structure emerged and eventually gave rise to an increase in the surface hardness.

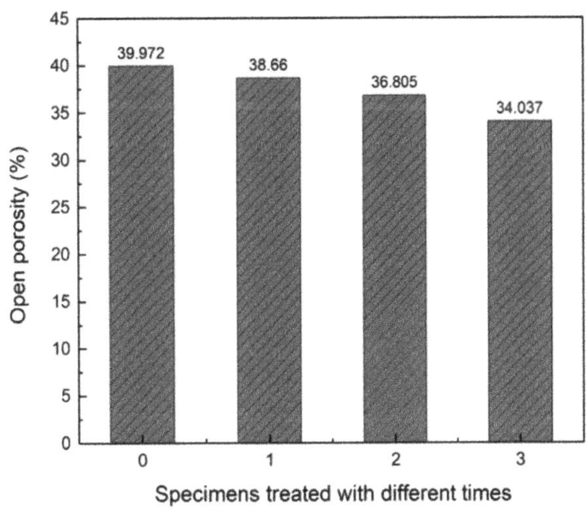

Figure 12. Open porosity of the specimens before and after treatment.

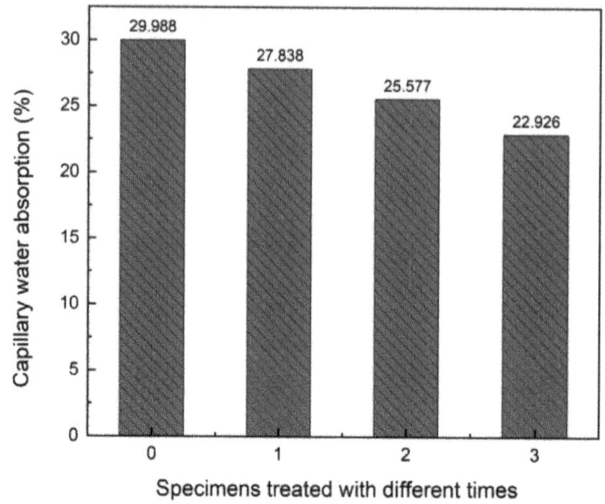

Figure 13. Capillary water absorption of the specimens before and after treatment.

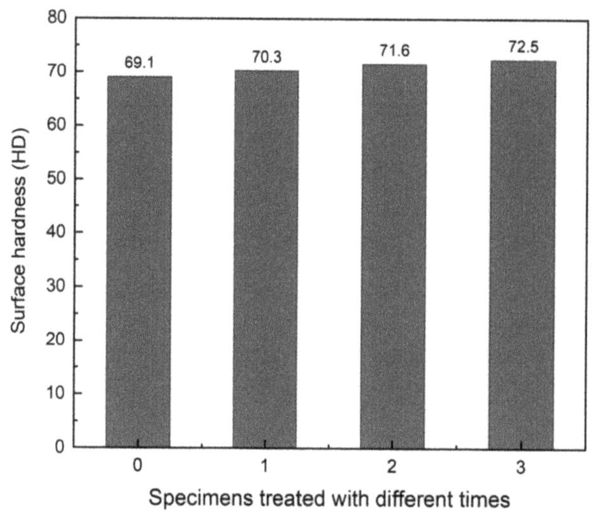

Figure 14. Surface hardness of the specimens before and after treatment.

4. Conclusions

Herein, an innovative route using a methanol solution of barium hydroxide-urea for the conservation of stone artifacts with gypsum weathering crusts was investigated. After the introduction of the barium hydroxide-urea methanol solution and water, in sequence, the barium sulfate and calcium carbonate layer were formed to provide prolonged protection for the brittle stone artifacts; this reaction can be completed within one week. Because of the low content of carbon dioxide in the natural atmosphere and the complete carbonization reaction difficulty that the internal substrate could face, the urea was added as a modifier into the methanol solution of barium hydroxide to provide a continuous source of carbonate ions. In addition, the existence of the urea did not affect the high solubility of the barium hydroxide in the methanol or the ability of the barium hydroxide methanol solution to penetrate deep into the substrate.

The conservation effect of the methanol solution of barium hydroxide-urea was satisfying. The conductivity and IC results reveal a clear increase in the water resistance ability of the treated specimens, which can be up to ten times stronger than the blank specimens. Meanwhile, the results of a larger surface hardness value indicate more solid structures of the surface gypsum crusts of the treated specimens. In addition, the results of the color difference, wetting angle, open porosity, and capillary water absorption denote that the inherent physical properties of the gypsum crust and original appearance were hardly altered. These positive results show that the methanol solution of barium hydroxide-urea could have great potential in the conservation of stone artifacts with gypsum weathering crusts.

However, there are still limitations to this method that can be further studied. It was noticed that the urea could produce carbon dioxide as well as ammonia, which is an irritant. Thus, how to reduce the irritation of ammonia still needs further investigation. In addition, the use of the methanol solvent could be detrimental to human health. Hence, there is a need to find ways to diminish this harm. We hope that through a continuous investigation, there will be more exciting findings about this novel conservation method.

Author Contributions: Conceptualization, F.Y.; Data curation, R.L., L.H., T.L., Y.L. and X.C.; Investigation, R.L., L.H., Y.L., T.L. and K.Z.; Methodology, F.Y.; Writing—original draft, R.L.; Writing—review & editing, R.L. and Y.L. All authors have read and agreed to the published version of the manuscript.

Funding: This research was funded by the National Natural Science Foundation of China (B0501 21975202, 52108031); the Collaborative Innovation Team Foundation of University in Gansu Province (2017C-20); 111 project (D18004) and the Key R&D Program in Shaanxi Province (2020SF-363).

Institutional Review Board Statement: Not applicable.

Informed Consent Statement: Not applicable.

Data Availability Statement: Not applicable.

Conflicts of Interest: The authors declare no conflict of interest.

References

1. Bozdağ, A.; Ince, I.; Bozdağ, A.; Hatır, M.E.; Tosunlar, M.B.; Korkanç, M. An assessment of deterioration in cultural heritage: The unique case of Eflatunpınar Hittite Water Monument in Konya, Turkey. *Bull. Eng. Geol. Environ.* **2020**, *79*, 1185–1197. [CrossRef]
2. Camuffo, D. Deterioration Processes of Historical Monuments. In *Studies Environmental Science*; Schneider, T., Ed.; Elsevier: Amsterdam, The Netherlands, 1986; Volume 30, pp. 189–221. [CrossRef]
3. Charola, A.E.; Ware, R. Acid deposition and the deterioration of stone: A brief review of a broad topic. *Geol. Soc. Lond. Spec. Publ.* **2002**, *205*, 393–406. [CrossRef]
4. Ausset, P.; Lefèvre, R.A.; Del Monte, M. Early mechanisms of development of sulphated black crusts on carbonate stone. In *Proceedings of the 9th International Congress on Deterioration and Conservation of Stone, Venice, Italy, 19–21 June 2000*; Fassina, V., Ed.; Elsevier Science B.V.: Amsterdam, The Netherlands, 2000; pp. 329–337. [CrossRef]
5. Dean, J.A. *Lange's Handbook of Chemistry*; Wei, J.F., Ed.; Science Press: Beijing, China, 2003.
6. Kramar, S.; Urosevic, M.; Pristacz, H.; Mirtič, B. Assessment of limestone deterioration due to salt formation by micro-Raman spectroscopy: Application to architectural heritage. *J. Raman Spectrosc.* **2010**, *41*, 1441–1448. [CrossRef]
7. Fronteau, G.; Schneider-Thomachot, C.; Chopin, E.; Barbin, V.; Mouze, D.; Pascal, A. Black-crust growth and interaction with underlying limestone microfacies. *Geol. Soc. Lond. Speec. Publ.* **2010**, *333*, 25–34. [CrossRef]
8. Zanini, A.; Trafeli, V.; Bartoli, L. The laser as a tool for the cleaning of Cultural Heritage. *IOP Conf. Ser. Mater. Sci. Eng.* **2018**, *364*, 012078. [CrossRef]
9. Gioventù, E.; Lorenzi, P. Bio-Removal of Black Crust from Marble Surface: Comparison with Traditional Methodologies and Application on a Sculpture from the Florence's English Cemetery. *Procedia Chem.* **2013**, *8*, 123–129. [CrossRef]
10. Pozo-Antonio, J.; Rivas, T.; López, A.; Fiorucci, M.; Ramil, A. Effectiveness of granite cleaning procedures in cultural heritage: A review. *Sci. Total Environ.* **2016**, *571*, 1017–1028. [CrossRef]
11. Liu, Y.; Dong, T.; Zhang, K.; Yang, F.; Wang, L. Preliminary Study of the Targeted Cleaning of an Artificial Gypsum Layer on White Marble. *Coatings* **2021**, *11*, 37. [CrossRef]
12. Zezza, F. The Monument Stone: An Eternal Link of Past Civilizations. In *10th International Symposium on the Conservation of Monuments in the Mediterranean Basin: Natural and Anthropogenic Hazards and Sustainable Preservation*; Koui, M., Zezza, F., Kouis, D., Eds.; Springer International Publishing: Cham, Switzerland, 2018; pp. 17–28. [CrossRef]

13. Artesani, A.; Di Turo, F.; Zucchelli, M.; Traviglia, A. Recent Advances in Protective Coatings for Cultural Heritage—An Overview. *Coatings* **2020**, *10*, 217. [CrossRef]
14. Zhang, H.; Liu, Q.; Liu, T.; Zhang, B. The preservation damage of hydrophobic polymer coating materials in conservation of stone relics. *Prog. Org. Coat.* **2013**, *76*, 1127–1134. [CrossRef]
15. Frigione, M.; Lettieri, M. Novel Attribute of Organic–Inorganic Hybrid Coatings for Protection and Preservation of Materials (Stone and Wood) Belonging to Cultural Heritage. *Coatings* **2018**, *8*, 319. [CrossRef]
16. Li, W.; Lin, J.; Zhao, Y.; Pan, Z. The Adverse Effects of TiO_2 Photocatalycity on Paraloid B72 Hybrid Stone Relics Protective Coating Aging Behaviors under UV Irradiation. *Polymers* **2021**, *13*, 262. [CrossRef] [PubMed]
17. Hansen, E.F.; Doehne, E.; Fidler, J.M.; Larson, J.D.; Martin, B.R.; Matteini, M.; Rodriguez-Navarro, C.; Pardo, E.S.; Price, C.; De Tagle, A.; et al. A review of selected inorganic consolidants and protective treatments for porous calcareous materials. *Stud. Conserv.* **2003**, *48*, 13–25. [CrossRef]
18. Drdácký, M.; Slížková, Z. Calcium hydroxide based consolidation of lime mortars and stone. In Proceedings of the International Symposium 'Stone Consolidation in Cultural Heritage', Lisbon, Portugal, 23–25 March 2022; pp. 299–308. Available online: https://www.researchgate.net/publication/313064616 (accessed on 31 October 2022).
19. Hall, L.R.; Matero, F.G. Considerations on Complex Sequential Treatments of Gypsum Crusts: The Carrara Marble Capitals of the Philadelphia Merchants' Exchange. *J. Am. Inst. Conserv.* **2011**, *50*, 123–148. [CrossRef]
20. Sayre, E.V. Direct Deposition of Barium Sulfate from Homogeneous Solution within Porous Stone. *Stud. Conserv.* **1971**, *16*, 115–117. [CrossRef]
21. Baglioni, P.; Chelazzi, D.; Giorgi, R. *Nanotechnologies in the Conservation of Cultural Heritage: A Compendium of Materials and Techniques*; Baglioni, P., Chelazzi, D., Giorgi, R., Eds.; Springer: Amsterdam, The Netherlands, 2015; pp. 15–59. [CrossRef]
22. Magrini, D.; Bartolozzi, G.; Bracci, S.; Carlesi, S.; Cucci, C.; Picollo, M. Evaluation of the efficacy and durability of the "barium hydroxide method" after 40 years. Multi-analytical survey on the Crocifissione by Beato Angelico. *J. Cult. Herit.* **2020**, *45*, 362–369. [CrossRef]
23. Toniolo, L.; Colombo, C.; Realini, M.; Peraio, A.; Positano, M. Evaluation of barium hydroxide treatment efficacy on a dolomitic marble. *Ann. Chim.* **2001**, *91*, 813–821.
24. Salvadori, B.; Errico, V.; Mauro, M.; Melnik, E.; Dei, L. Evaluation of Gypsum and Calcium Oxalates in Deteriorated Mural Paintings by Quantitative FTIR Spectroscopy. *Spectrosc. Lett.* **2003**, *36*, 501–513. [CrossRef]
25. Pastuović, Ž.; Fazinić, S.; Jakšić, M.; Krstić, D.; Mudronja, D. The use of the RBI nuclear microprobe in conservation process studies of a church portal. *Nucl. Instrum. Methods Phys. Res. Sect. B Beam Interact. Mater. Atoms* **2005**, *231*, 546–552. [CrossRef]
26. Sassoni, E.; Graziani, G.; Franzoni, E.; Scherer, G.W. Conversion of calcium sulfate dihydrate into calcium phosphates as a route for conservation of gypsum stuccoes and sulfated marble. *Constr. Build. Mater.* **2018**, *170*, 290–301. [CrossRef]
27. Chelazzi, D.; Poggi, G.; Jaidar, Y.; Toccafondi, N.; Giorgi, R.; Baglioni, P. Hydroxide nanoparticles for cultural heritage: Consolidation and protection of wall paintings and carbonate materials. *J. Colloid Interface Sci.* **2013**, *392*, 42–49. [CrossRef] [PubMed]
28. Saoud, K.M.; Ibala, I.; El Ladki, D.; Ezzeldeen, O.; Saeed, S. *Digital Heritage. Progress in Cultural Heritage: Documentation, Preservation, and Protection*; Ioannides, M., Ed.; Springer International Publishing: Berlin/Heidelberg, Germany, 2014; pp. 342–352. [CrossRef]
29. Lu, R.; Wang, L.; Liu, Y.; Yang, F.; Yang, L.; Wang, L.; Gao, X. Conservation of surface gypsification stone relics using methanol solution of barium hydroxide as a novel treating agent. *Appl. Phys. A* **2021**, *128*, 37. [CrossRef]
30. Durán-Suárez, J.; García-Beltrán, A.; Rodríguez-Gordillo, J. Colorimetric cataloguing of stone materials (biocalcarenite) and evaluation of the chromatic effects of different restoring agents. *Sci. Total Environ.* **1995**, *167*, 171–180. [CrossRef]
31. Camuffo, D.; Del Monte, M.; Sabbioni, C.; Vittori, O. Wetting, deterioration and visual features of stone surfaces in an urban area. *Atmos. Environ.* **1982**, *16*, 2253–2259. [CrossRef]
32. Cabañas, M.V.; Rodríguez-Lorenzo, L.M.; Vallet-Regí, M. Setting Behavior and in Vitro Bioactivity of Hydroxyapatite/Calcium Sulfate Cements. *Chem. Mater.* **2002**, *14*, 3550–3555. [CrossRef]
33. Favaro, M.; Tomasin, P.; Ossola, F.; Vigato, P.A. A novel approach to consolidation of historical limestone: The calcium alkoxides. *Appl. Organomet. Chem.* **2008**, *22*, 698–704. [CrossRef]
34. Rodriguez-Navarro, C.; Suzuki, A.; Ruiz-Agudo, E. Alcohol dispersions of calcium hydroxide nanoparticles for stone conservation. *Langmuir* **2013**, *29*, 11457–11470. [CrossRef] [PubMed]
35. Kroftová, K.; Škoda, D.; Kuřitka, I.; Kubát, J. Technology of preparation of barium and magnesium hydroxide nanodispersion and possibilities of their use in monument care. *Contem. Mater. Technol. Civ. Eng.* **2018**, *2019*, 21–23. [CrossRef]
36. Starikova, Z.A.; Kessler, V.G.; Turova, N.Y.; Dantsker, I.A.; Bobyljov, A.P.; Mitiaev, A.S. Interaction of barium oxide and hydroxide with methanol: X-ray single crystal study of Ba(OH)2 methanol solvates. *Polyhedron* **2006**, *25*, 2401–2406. [CrossRef]
37. Hussain, R.; Devi, R.R.; Maji, T.K. Controlled release of urea from chitosan microspheres prepared by emulsification and cross-linking method. *Iran. Polym. J.* **2012**, *21*, 473–479. [CrossRef]
38. Chakrabarty, D.; Mahapatra, S. Aragonite crystals with unconventional morphologies. *J. Mater. Chem.* **1999**, *9*, 2953–2957. [CrossRef]
39. Al Omari, M.M.H.; Rashid, I.S.; Qinna, N.A.; Jaber, A.M.; Badwan, A.A. Calcium Carbonate. In *Profiles of Drug Substances, Excipients and Related Methodology*; Elsevier: Amsterdam, The Netherlands, 2016; Volume 41, pp. 31–132.

40. Zuena, M.; Zendri, E.; Costa, D.; Delgado-Rodrigues, J.; El Habra, N.; Tomasin, P. Calcium Ethoxide as Consolidant for Porous Limestones: Influence of the Solvent. *Coatings* **2019**, *9*, 83. [CrossRef]
41. Ramaswamy, V.; Vimalathithan, R.M.; Ponnusamy, V. Synthesis and characterization of BaSO4 nano particles using micro emulsion technique. *Adv. Appl. Sci. Res.* **2010**, *1*, 197–204.
42. Tomasin, P.; Mondin, G.; Zuena, M.; El Habra, N.; Nodari, L.; Moretto, L.M. Calcium alkoxides for stone consolidation: Investigating the carbonation process. *Powder Technol.* **2019**, *344*, 260–269. [CrossRef]
43. Wilcox, L.V. Electrical Conductivity. *J. AWWA* **1950**, *42*, 775–776. [CrossRef]
44. Sassoni, E.; Graziani, G.; Franzoni, E. An innovative phosphate-based consolidant for limestone. Part 1: Effectiveness and compatibility in comparison with ethyl silicate. *Constr. Build. Mater.* **2016**, *102*, 918–930. [CrossRef]
45. Liu, Q.; Zhang, B.J. Assessment of Damage from Organic Protective Coating Treatments to Historic Stone Buildings and Sculptures. *Appl. Mech. Mater.* **2010**, *44–47*, 610–613. [CrossRef]
46. García, O.; Malaga, K. Definition of the procedure to determine the suitability and durability of an anti-graffiti product for application on cultural heritage porous materials. *J. Cult. Herit.* **2012**, *13*, 77–82. [CrossRef]
47. Tsakalof, A.; Manoudis, P.; Karapanagiotis, I.; Chryssoulakis, I.; Panayiotou, C. Assessment of synthetic polymeric coatings for the protection and preservation of stone monuments. *J. Cult. Herit.* **2007**, *8*, 69–72. [CrossRef]

coatings

MDPI

Article

One-Step Preparation of High Performance TiO_2/CNT/CQD Nanocomposites Bactericidal Coating with Ultrasonic Radiation

Jin Xiang [1,2], Shuchang Wang [3], Yuanxin Cao [3], Lining Fang [4], Wei Ke [3], Hui Guo [3,*], Baoyu Duan [5,*], Wenhe Yu [6], Liang Li [7] and Zilong Zhao [3,*]

1 Department of Mechanics, Jinzhong University, Jinzhong 030619, China
2 Leo Group Pump (Zhejiang) Co., Ltd., Taizhou 318000, China
3 School of Chemical Engineering and Technology, Sun Yat-sen University, Zhuhai 519082, China
4 Department of Environmental Science, Hebei University of Environmental Engineering, Qinghuangdao 066000, China
5 College of Locomotive and Rolling Stock Engineering, Dalian Jiaotong University, Dalian 116028, China
6 Beijing Key Laboratory of Materials Utilization of Nonmetallic Minerals and Solid Wastes, National Laboratory of Mineral Materials, School of Materials Science and Technology, China University of Geosciences, Beijing 100083, China
7 School of Physics, Engineering and Computer Science, University of Hertfordshire, Hatfield AL10 9AB, UK
* Correspondence: guoh37@mail.sysu.edu.cn (H.G.); duanby83@163.com (B.D.); zhaozlong@mail.sysu.edu.cn (Z.Z.)

Abstract: As an environmental semiconductor material, TiO_2 has important applications in the fields of environmental protection and water treatment. The preparation of P25 particles into nano-functional material films with a high specific surface area has always been a bottleneck limiting its large-scale application. In this paper, a one-step method of preparing TiO_2 nanocomposites by doping carbon nanotube (CNT) and carbon quantum dots (CQD) with tetrabutyltitanate and P25 TiO_2 under ultrasonic radiation is proposed to synthesize a novel antifouling material, which both eliminates the bacterium of Escherichia coli and shows good photoelectric properties, indicating a great value for the industrial promotion of TiO_2/CNT. This mesoporous composite exhibits a high specific surface area of 78.07 M^2/g (BET) and a tested pore width range within 10–120 nm. The surface morphology of this composite is characterized by TEM and the microstructure is characterized through XRD. This preparation method can fabricate P25 particles into a nano-functional material film with a high specific surface area at a very low cost.

Keywords: TiO_2; carbon quantum dots; one-step synthesis; ultrasonic radiation; carbon nanotube; mesoporous nanocomposites

Citation: Xiang, J.; Wang, S.; Cao, Y.; Fang, L.; Ke, W.; Guo, H.; Duan, B.; Yu, W.; Li, L.; Zhao, Z. One-Step Preparation of High Performance TiO_2/CNT/CQD Nanocomposites Bactericidal Coating with Ultrasonic Radiation. *Coatings* **2023**, *13*, 145. https://doi.org/10.3390/coatings13010145

Academic Editor: Joaquim Carneiro

Received: 27 November 2022
Revised: 2 January 2023
Accepted: 3 January 2023
Published: 11 January 2023

1. Introduction

Photosynthesis in nature achieves a highly efficient solar energy conversion mainly through combining several different molecules to arrange the nanoscale, which indicates the importance of a technique for the site-selective coupling of different materials to realize artificially efficient devices [1–3]. TiO_2 is a kind of safe, stable, and cheap material, which has attracted significant research interest [4–9]. Photo-catalytic reactions on the semi-conductive surfaces of TiO_2 materials are widely investigated and reported for the purposes of (1) water splitting to produce H_2 and (2) eliminating pollutants from water and air [10–15]. Obviously, the direct splitting of water molecules through Photo-electrochemical (PEC) processes is a sustainable green approach to convert water and sunlight to hydrogen and oxygen to achieve [16,17]. Much effort has been made in the selection of efficient stable semiconductor materials to build PEC cells, thus bringing silicon (Si), III–V compounds, and various oxides into sight [18–23]. In the first-row of transition metal oxides, TiO_2 is a poor water oxidation catalyst with a large over potential for the sluggish kinetics and oxygen evolution reaction [14]. Hence, it is essential to develop a TiO_2-based visible light responsive photo-catalyst, and much work has been done to develop "second-generation" TiO_2 and other narrow band gap semiconductors to absorb visible light [17,24–27]. Though nano-sized particles could be prepared by the sol-gel method for the film's preparation, the size of the particles is difficult to control [28–32]. The introduction of the graphene/graphene derivatives (including graphene oxide (GO) and other forms of functionalized graphene), the 3D single atom thick carbon layer, as the framework of the membrane, enables the morphology of the nanostructured materials to be extended from 1D to 0D (nanoparticles). The range of the nanostructured materials chosen is extended from only the photocatalytic materials (TiO_2, ZnO, CoS, CdS, CdSe, MoO_x, etc.) to other functional materials (MnO_2 for oxidation, Ag for disinfection, Pt for catalysis, etc.). Applying the 3D graphene-based nanocomposite multicomposite membrane with an expected thickness < 100 nm for the desalination applications will potentially create a new and promising industrial sector, and these applications are still undeveloped.

It is known that TiO_2 is a wide bandgap semiconductor, with a low-to-moderate response rate to sunlight, and research has focused on reducing the absorption edge energy to improve the utilization of sunlight [33,34]. Moreover, the large specific area of mesoporous TiO_2 nanoparticles is another favorable factor for generating large quantities of active sites for a photocatalytic reaction, which has taken place on the surface of the photocatalyst [35]. Additionally, the preparation of P25 particles into nano-functional material films with a high specific surface area has always been a bottleneck limiting its large-scale application.

In this work, sol was prepared with tetrabutyl titanate; commercial P25 TiO_2 was added to ensure the uniformity of the nanoparticles; and CNT and CQD were doped to improve the light absorptivity of the semi-conductive materials. CQD also contributed to the gel's formation.

2. Experimental Section

2.1. Synthesis

$Ti(O\text{-}i\text{-}C_4H_9)_4$ (TB, Aladdin, AR) (≥98% Ti) and ethanol absolute (EtOH, Aladdin, Nanjing, China, AR) were blended with a 5% (wt.) multi-wall carbon nanotube (Nano-Com Shanghai, China,) average particle size 10–20 nm) to make a mixture. Hydrochloride acid was added as a catalyst to enhance the hydrolysis rate. Commercial P25 TiO_2 (Innochem) particles with an average particle size of 25 nm were dispersed into the Ti-based sol at 0.2 g/mL in vigorous magnetic stirring for 30 min. GQD nanomaterials with a diameter of 15 nm were purchased from Aladdin (G196610-100 mL). GQDs were dispersed into the tetra-n-butyl orthotitanate at 80 mg/mL in vigorous magnetic stirring for more than 5 min. The mixture was then stirred via vigorous mechanical stirring and subsequently placed in an 80 °C water bath with an ultrasonic (Dakou, Kongshan, China, KQ-300VDE, 45 kHz) for 5 h. After that, the sonicated mixture was dried overnight at a temperature of 80 °C and then calcined in an Ar atmosphere at 450 °C for 6 h before the grounding procedure to obtain an anatase phase. The schematic of the formation of the mesoporous TiO_2/CNT/CQD nanocomposites membrane is shown in Scheme 1.

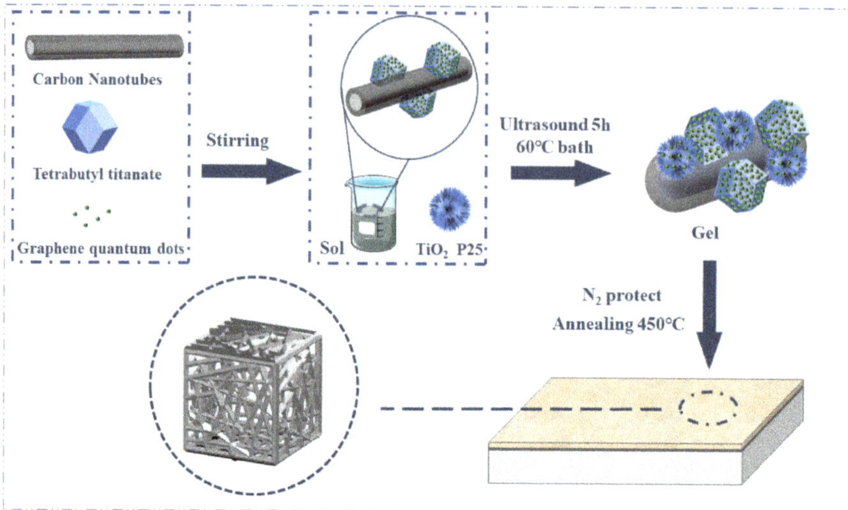

Scheme 1. Formation of mesoporous TiO_2/CNT/CQD nanocomposites membrane by sol-gel ultrasonic irradiation method.

2.2. Characterization

Various characterizing techniques, including a transmission electron microscope (FEI Talos-F200S TEM), pore structure determination (Brunauer–Emmett–Teller, BET method), and X-ray diffraction (XRD), were applied to analyze the material. A transmission electron microscope (JEOL, JEOL-2100F, Tokyo, Japan) was used to characterize the microstructure. UV–vis DRS was performed on a spectrophotometer (Lambda750, PerkinElmer, Inc., Waltham, MA, USA). The absorbance was tested using a UV–vis spectro-photometer (722, Weimipai Technology Co., Ltd., Hangzhou, China). The light absorption properties were measured using a Fourier Infrared Instrument (PerkinElmer Spectrum, Akron, OH, USA). The absorbance properties were measured using the Zolix SS150 (Zolix Instruments Co., Ltd., Beijing, China). The light intensity and photoluminescence spectrum's structure were determined using a fluorescence spectrometer (F-4500, Hitachi, Ibaraki, Japan). The photoelectrochemical properties were measured on an electrochemical workstation (Chenhua, CHI660E, Shanghai, China).

2.3. Antibacterial Tests

To test the materials with the bacterium of Escherichia coli, typically, the sample (10 mg) was added to 10 mL of pure water to make a TiO_2 solution (solution T) with TiO_2 suspended within. To remove the microorganisms from the system, the whole set was sterilized in an automatic high-temperature sterilization pot. Then, using a conventional bacterial solution with its original concentrate of 1.15 g/L, a 0.01% diluted solution was prepared and 100 µL of this solution was added into the TiO_2 solution (solution T) as a source of Escherichia coli. The mixture was irritated under 34.4 Klux solar light for 1 h. After lighting, the antibacterial solution was coated on the prepared solid medium (10 g of pancreatic protein, 5 g of yeast powder, 10 g of NaCl, and 20 g of agar diluted with 1 L of pure water) and was tested after 8 h for its anti-bacterial performance.

3. Results and Discussion

Figure 1 presents the results of the BET test of the TiO_2/CNT/CQD nanocomposites. The surface area of TiO_2/CNT/CQD is as high as 78.07 m^2/g and the pore width range is 10–120 nm, indicating that an ultrasonic radiation treatment is a useful technique to prepare mesoporous materials [36–39]. Bai has reported that the surface area of the TiO_2/CNT mesoporous composite material is 42.90 m^2/g [40]. Therefore, the doping of CQD is beneficial to increase the surface of the TiO_2/CNT/CQD nanocomposites. It attributes to the small particle size.

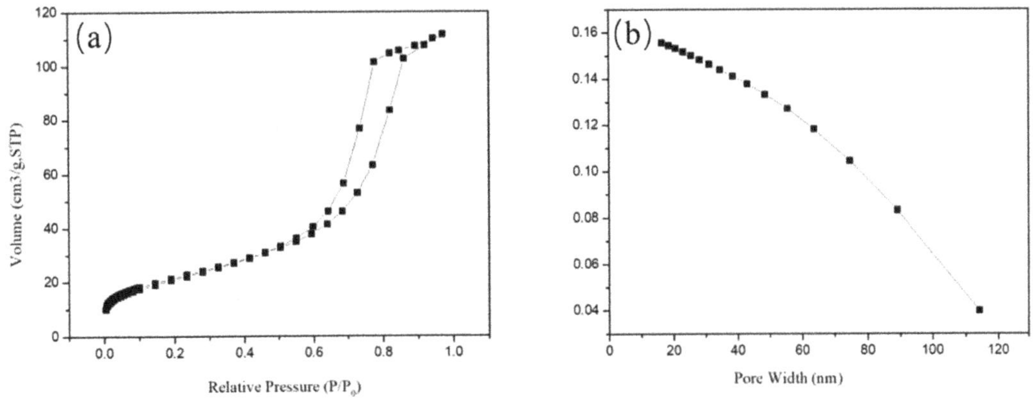

Figure 1. Isotherms of nitrogen adsorption–desorption and pore width of the TiO_2/CNT/CQD mesoporous composite. (**a**) Nitrogen adsorption–desorption isotherms and pore width of TiO_2/CNT/CQD, (**b**) aperture distribution map.

Figure 2 shows the energy disperse spectroscopy (EDS) mapping results of the TiO_2/CNT/CQD mesoporous nanocomposites prepared under ultrasonic radiation. Small TiO_2 particles appear to cluster around the carbon nanotube, indicating that the nanotube is a sufficient support material for a better TiO_2/CNT dispersion in this specific preparation. Although the composite interface between TiO_2 and the carbon nanotube, which appears to be a loose one, needs to be further improved, the doping of TiO_2 with CNT and CQD obviously improves the dispersion of TiO_2 as well as promotes the formation of hetero junctions at the TiO_2-CNT interface, which favors the improvement of the absorption efficiency of visible light [41,42].

The TEM results of TiO_2/CNT/CQD prepared with the ultrasonic treatment are shown in Figure 3. Figure 3a depicts the agglomeration of small particles in the TiO_2/CNT/CQD mesoporous nanocomposites. In Figure 3b, the interface between the two main phases, namely the TiO_2 and the CNT, has been remarkably enhanced through a high temperature. The high-resolution diagrams, Figure 3c,d, correspond to anatase TiO_2 [43,44], which clearly represent the relevant information at the atomic scale. Figure 3e,f are high-resolution TEM diagrams of CQD. From Figure 4, it can be seen that the morphology and spacing between the crystal planes are CQD, and CQD is successfully doped into P25 and CNT by sol-gel, but there is only a small amount of reunion.

The FTIR spectra have indicated that the characteristic bands at 3400 cm^{-1} and 1630 cm^{-1} correspond to the surface water and hydroxyl group [45]. Figure 4 shows the XRD results of the TiO_2/CNT/CQD mesoporous composite. As shown in Figure 4, it can be found that the TiO_2/CNT/CQD mesoporous composite is an obvious anatase phase and C phase, which is conducive to improving the photoelectric properties of the material. TiO_2 nanoparticles have attracted great interest because of their special physical and chemical properties, especially as photocatalytic oxidation catalysts in corresponding device materials and environmental pollution control [46].

Marine fouling is a critical issue in modern marine science and technology, which strictly determines marine transportation and farming. The existing antifouling approaches, such as antifouling hydrogel coatings and organic tin coating [47–49], always demonstrate an unsatisfactory performance. In our system, the material shows the CV curves and anti-bacterial properties of the control, TiO_2/CNT and TiO_2/CNT/CQD, using Escherichia coli bacteria (Figure 5). Comparing the bacteriostasis processes uses TiO_2/CNT (Figure 5c) and TiO_2/CNT/CQD (Figure 5d) as anti-bacterium agents, respectively. In this experiment, obviously, the number of Escherichia coli bacterial colonies in Petri dishes all increased with time. The sample (c) with TiO_2/CNT/CQD as an anti-bacterium agent got the bacterium of Escherichia coli under control, while the sample using TiO_2 as an anti-bacterium agent just grew more bacterial colonies, indicating that doping TiO_2 with CNT and CQD contributes to boosting the TiO_2s performance in an environmental pollution control.

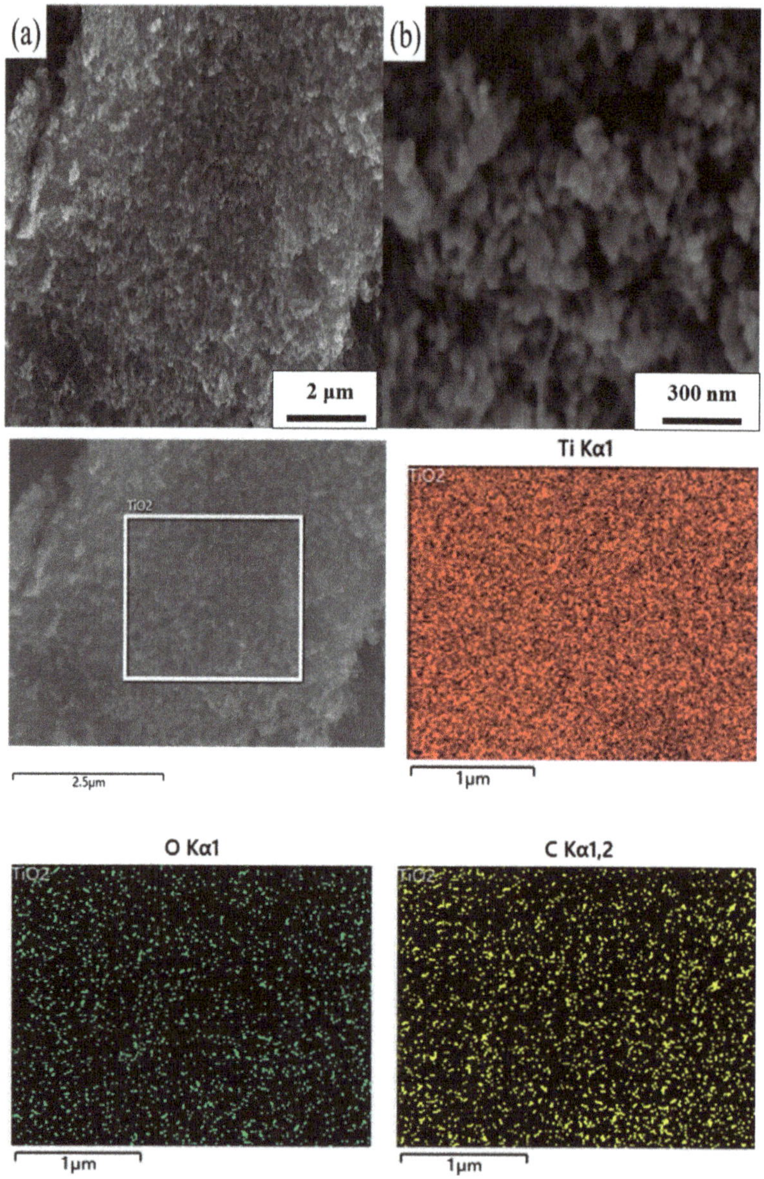

Figure 2. EDS mapping of TiO_2/CNT/CQD: (**a**,**b**) are SEM images of the TiO_2/CNT/CQD mesoporous composite.

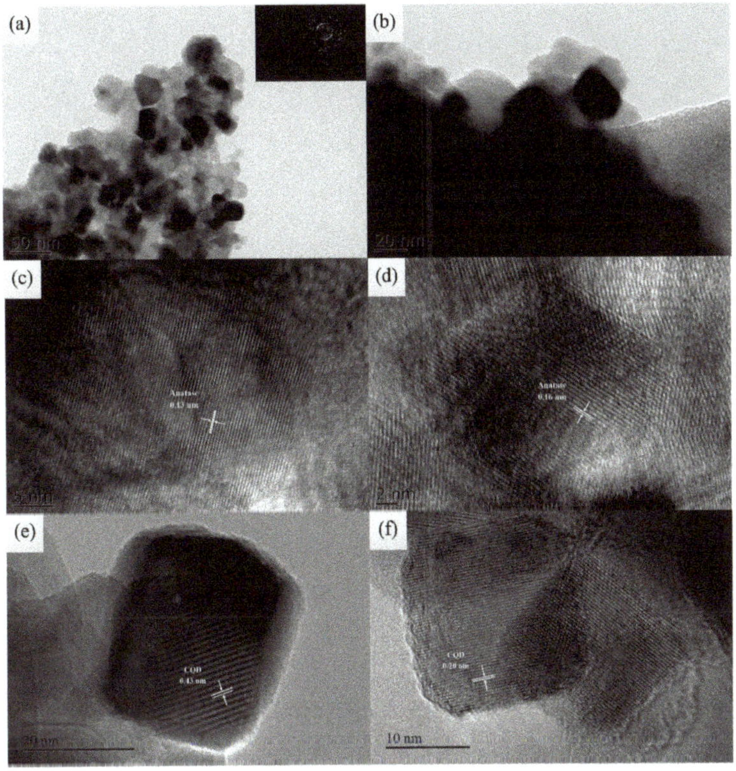

Figure 3. TEM images of the TiO_2/CNT/CQD mesoporous composite: (**a**,**b**) are TEM images of the TiO_2/CNT/CQD mesoporous composite, (**c**,**d**) are HRTEM images of the anatase phase, (**e**,**f**) HRTEM is images of the CQD phase.

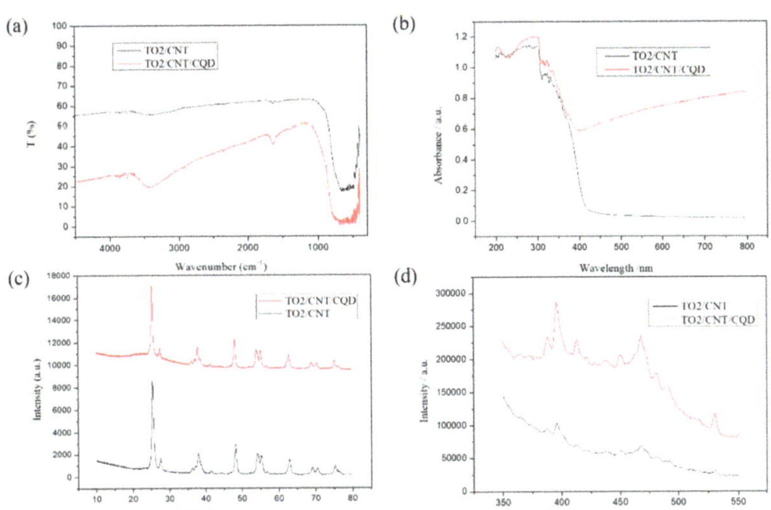

Figure 4. (**a**) FT-IR spectra, (**b**) UV–vis, (**c**) XRD patterns, and (**d**) photoluminescence spectra of TiO_2/CNT and TiO_2/CNT/CQD mesoporous composites.

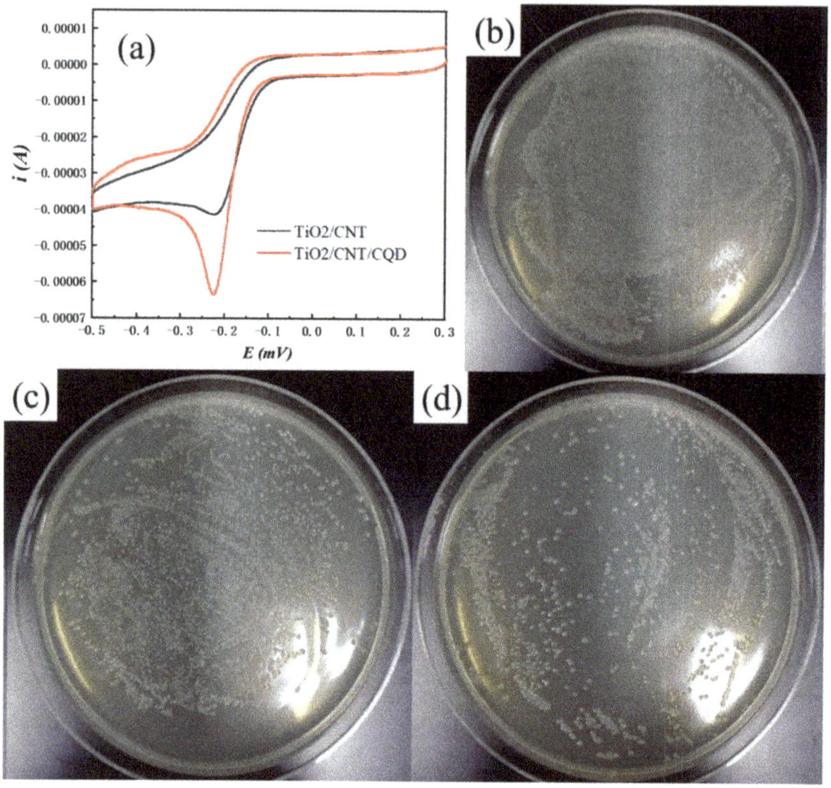

Figure 5. (**a**) CV curves and anti-bacterial properties of (**b**) control, (**c**) TiO_2/CNT, and (**d**) TiO_2/CNT/CQD using Escherichia coli.

4. Conclusions

The main conclusions were as follows:

(1) The composites prepared exhibit a large specific surface area of 78.07 m^2/g and a pore width of 10–120 nm, indicating that an ultrasonic radiation treatment contributes to forming nanocomposites with a high specific surface area.
(2) Commercial P25 particles can be prepared into gel films at a low cost by sol-gel ultrasonic radiation.
(3) TiO_2/CNT/CQD nanocomposites prepared by doping CNT/CQD can significantly improve the visible light absorption efficiency and bactericidal efficiency.

Author Contributions: Conceptualization, H.G., B.D. and Z.Z.; methodology, S.W. and Y.C.; validation, W.K. and L.F.; formal analysis, J.X., L.L. and W.Y.; investigation, J.X. and Z.Z.; writing—original draft preparation, J.X. and H.G.; writing—review and editing, J.X. and Z.Z. All authors have read and agreed to the published version of the manuscript.

Funding: This research received no external funding.

Institutional Review Board Statement: Not applicable.

Informed Consent Statement: Not applicable.

Data Availability Statement: Not applicable.

Conflicts of Interest: The authors declare no conflict of interest.

References

1. Tada, H.; Mitsui, T.; Kiyonaga, T.; Akita, T.; Tanaka, K. All-solid-state Z-scheme in CdS-Au-TiO_2 three-component nanojunction system. *Nat. Mater.* **2006**, *5*, 782–786. [CrossRef] [PubMed]
2. Geng, W.; Wang, L.; Yang, X.-Y. Nanocell hybrids for green chemistry. *Trends Biotechnol.* **2022**, *40*, 974–986. [CrossRef] [PubMed]
3. Li, H.L.; Luo, S.W.; Zhang, L.Q.; Zhao, Z.L.; Wu, M.; Li, W.H.; Liu, F.Q. Water- and Acid-Sensitive Cu_2O@Cu-MOF Nano Sustained-Release Capsules with Superior Antifouling Behaviors. *ACS Appl. Mater. Interfaces* **2022**, *14*, 1910–1920. [CrossRef] [PubMed]
4. Zhao, L.; Guan, B.; Xin, Y.; Huang, X.; Liu, C.; Wu, P.; Liu, Q. A quantitative study on mechanical behavior of Mg alloys with bimodal texture components. *Acta Mater.* **2021**, *214*, 117013. [CrossRef]
5. Wu, T.W.; Kong, W.H.; Zhang, Y.; Xing, Z.; Zhao, J.X.; Wang, T.; Shi, X.F.; Luo, Y.L.; Sun, X.P. Greatly Enhanced Electrocatalytic N-2 Reduction on TiO_2 via V Doping. *Small Methods* **2019**, *3*, 8. [CrossRef]
6. Wang, Y.X.; Rao, L.; Wang, P.F.; Shi, Z.Y.; Zhang, L.X. Photocatalytic activity of N-TiO_2/O-doped N vacancy g-C_3N_4 and the intermediates toxicity evaluation under tetracycline hydrochloride and Cr(VI) coexistence environment. *Appl. Catal. B-Environ.* **2020**, *262*, 12. [CrossRef]
7. Zhao, Z.; Lai, H.S.; Li, H.; Li, L. Preparation and Properties of Graphene Doped TiO_2 Mesoporous Materials for Photocathode Protection. *Int. J. Electrochem. Sci.* **2021**, *16*, 210316. [CrossRef]
8. Zhao, Z.; Jiang, X.; Li, S.; Li, L.; Feng, Z.; Lai, H. Microstructure Characterization and Battery Performance Comparison of MOF-235 and TiO_2-P25 Materials. *Crystals* **2022**, *12*, 152. [CrossRef]
9. Feng, Z.Y.; Hurley, B.; Zhu, M.L.; Yang, Z.; Hwang, J.; Buchheit, R. Corrosion Inhibition of AZ31 Mg Alloy by Aqueous Selenite (SeO_3^{2-}). *J. Electrochem. Soc.* **2019**, *166*, C520–C529. [CrossRef]
10. Tripathy, J.; Lee, K.; Schmuki, P. Tuning the Selectivity of Photocatalytic Synthetic Reactions Using Modified TiO_2 Nanotubes. *Angew. Chem.-Int. Ed.* **2014**, *53*, 12605–12608. [CrossRef]
11. Kong, J.J.; Rui, Z.B.; Ji, H.B. Carbon Nitride Polymer Sensitization and Nitrogen Doping of SrTiO3/TiO_2 Nanotube Heterostructure toward High Visible Light Photocatalytic Performance. *Ind. Eng. Chem. Res.* **2017**, *56*, 9999–10008. [CrossRef]
12. Kong, J.J.; Lai, X.D.; Rui, Z.B.; Ji, H.B.; Ji, S.F. Multichannel charge separation promoted ZnO/P25 heterojunctions for the photocatalytic oxidation of toluene. *Chin. J. Catal.* **2016**, *37*, 869–877. [CrossRef]
13. Xiao, Y.-X.; Ying, J.; Chen, J.-B.; Dong, Y.; Yang, X.; Tian, G.; Wu, J.; Janiak, C.; Ozoemena, K.I.; Yang, X.-Y. Confined Ultrafine Pt in Porous Carbon Fibers and Their N-Enhanced Heavy d-π Effect. *Chem. Mater.* **2022**, *34*, 3705–3714. [CrossRef]
14. Nie, X.; Wang, J.; Duan, W.; Zhao, Z.; Li, L.; Zhang, Z. One-step preparation of C-doped TiO_2 nanotubes with enhanced photocatalytic activity by a water-assisted method. *Crystengcomm* **2021**, *23*, 3015–3025. [CrossRef]
15. Feng, Z.; Xu, C.C.; Zhang, D.; Buchheit, R. Corrosion Protective Film Formation on Mg Alloy AZ31 by Exposure to Dilute Selenite Solutions. *Materials* **2021**, *14*, 286. [CrossRef]
16. Zhao, Z.L.; Li, H.L.; Huang, L.X.; Tan, Y.; Liu, F.Q.; Li, W.H. Preparation of graphene quantum dots-doped TiO_2 nanocomposites via a sol-gel method for photocathodic protection. *Mol. Cryst. Liq. Cryst.* **2021**, *731*, 80–87. [CrossRef]
17. Asahi, R.; Morikawa, T.; Ohwaki, T.; Aoki, K.; Taga, Y. Visible-light photocatalysis in nitrogen-doped titanium oxides. *Science* **2001**, *293*, 269–271. [CrossRef]
18. Wang, C.J.; Zhao, Y.L.; Xu, H.; Li, Y.F.; Wei, Y.C.; Liu, J.; Zhao, Z. Efficient Z-scheme photocatalysts of ultrathin g-C_3N_4-wrapped Au/TiO_2-nanocrystals for enhanced visible-light-driven conversion of CO_2 with H_2O. *Appl. Catal. B-Environ.* **2020**, *263*, 13. [CrossRef]
19. Dong, F.; Guo, S.; Wang, H.; Li, X.; Wu, Z. Enhancement of the Visible Light Photocatalytic Activity of C-Doped TiO_2 Nanomaterials Prepared by a Green Synthetic Approach. *J. Phys. Chem. C* **2011**, *115*, 13285–13292. [CrossRef]
20. Huang, W.; Wang, X.; Xue, Y.; Yang, Y.; Ao, X. Hybrid nanostructures of mixed-phase TiO_2 for enhanced photoelectrochemical water splitting. *RSC Adv.* **2015**, *5*, 56098–56102. [CrossRef]
21. Hwang, S.H.; Kim, C.; Jang, J. SnO_2 nanoparticle embedded TiO_2 nanofibers-Highly efficient photocatalyst for the degradation of rhodamine B. *Catal. Commun.* **2011**, *12*, 1037–1041. [CrossRef]
22. Mei, B.; Pedersen, T.; Malacrida, P.; Bae, D.; Frydendal, R.; Hansen, O.; Vesborg, P.C.K.; Seger, B.; Chorkendorff, I. Crystalline TiO_2: A Generic and Effective Electron-Conducting Protection Layer for Photoanodes and -cathodes. *J. Phys. Chem. C* **2015**, *119*, 15019–15027. [CrossRef]
23. Tang, C.; Bai, H.; Liu, L.; Zan, X.; Gao, P.; Sun, D.D.; Yan, W. A green approach assembled multifunctional Ag/AgBr/TNF membrane for clean water production & disinfection of bacteria through utilizing visible light. *Appl. Catal. B-Environ.* **2016**, *196*, 57–67. [CrossRef]
24. Yu, W.; Zheng, B.; Mao, K.; Jiang, J.; Luo, B.; Wu, X.; Tao, T.; Min, X.; Mi, R.; Huang, Z.; et al. Interfacial structure and photocatalytic degradation performance of graphene oxide bridged chitin-modified TiO_2/carbon fiber composites. *J. Clean. Prod.* **2022**, *361*, 132261. [CrossRef]
25. Wang, J.; Nie, X.; Wang, W.; Zhao, Z.; Li, L.; Zhang, Z. Single-layer graphene-TiO_2 nanotubes array heterojunction as photoanode to enhance the photoelectric of DSSCs. *Optik* **2021**, *242*, 167245. [CrossRef]
26. Feng, Z.; Li, J.; Ma, J.; Su, Y.; Zheng, X.; Mao, Y.; Zhao, Z. EBSD Characterization of 7075 Aluminum Alloy and Its Corrosion Behaviors in SRB Marine Environment. *J. Mar. Sci. Eng.* **2022**, *10*, 740. [CrossRef]

27. Ying, J.; Janiak, C.; Xiao, Y.-X.; Wei, H.; Yang, X.-Y.; Su, B.-L. Shape-Controlled Surface-Coating to Pd@Mesoporous Silica Core–Shell Nanocatalysts with High Catalytic Activity and Stability. *Chem. Asian J.* **2018**, *13*, 31–34. [CrossRef]
28. Shen, L.; Ying, J.; Tian, G.; Jia, M.; Yang, X.-Y. Ultralong PtPd Alloyed Nanowires Anchored on Graphene for Efficient Methanol Oxidation Reaction. *Chem.—Asian J.* **2021**, *16*, 1130–1137. [CrossRef]
29. Xiao, Y.-X.; Ying, J.; Tian, G.; Tao, Y.; Wei, H.; Fan, S.-Y.; Sun, Z.-H.; Zou, W.-J.; Hu, J.; Chang, G.-G.; et al. Highly dispersed PtPd on graphitic nanofibers and its heavy d-π effect. *Appl. Catal. B Environ.* **2019**, *259*, 118080. [CrossRef]
30. Xiao, Y.X.; Ying, J.; Tian, G.; Zhang, X.Q.; Janiak, C.; Ozoemena, K.I.; Yang, X.Y. PtPd hollow nanocubes with enhanced alloy effect and active facets for efficient methanol oxidation reaction. *Chem. Commun.* **2021**, *57*, 986–989. [CrossRef]
31. Li, J.; Li, T.; Zeng, Y.; Chen, C.; Guo, H.; Lei, B.; Zhang, P.; Feng, Z.; Meng, G. A novel sol-gel coating via catechol/lysine polymerization for long-lasting corrosion protection of Mg alloy AZ31. *Colloids Surf. A Physicochem. Eng. Asp.* **2023**, *656*, 130361. [CrossRef]
32. Zheng, H.Y.; Guo, Y.C.; Shen, F.M. Production of titanium powder by metallothermic reduction of TiO_2 in cold pressed pellets. *J. Cent. South Univ.* **2021**, *28*, 48–57. [CrossRef]
33. Xue, S.G.; Tang, L.; Tang, T.; Zhang, F.; Lyu, H.G.; Liu, H.Y.; Jiang, J.; Huang, Y.H. Identifying the active sites in C-N codoped TiO_2 electrode for electrocatalytic water oxidation to produce H2O2. *J. Cent. South Univ.* **2022**, *29*, 3016–3029. [CrossRef]
34. Fan, B.Y.; Liu, H.B.; Wang, Z.H.; Zhao, Y.W.; Yang, S.; Lyu, S.Y.; Xing, A.; Zhang, J.; Li, H.; Liu, X.Y. Ferroelectric polarization-enhanced photocatalytic performance of heterostructured BaTiO3@TiO_2 via interface engineering. *J. Cent. South Univ.* **2021**, *28*, 3778–3789. [CrossRef]
35. Nong, S.; Dong, W.; Yin, J.; Dong, B.; Lu, Y.; Yuan, X.; Wang, X.; Bu, K.; Chen, M.; Jiang, S.; et al. Well-Dispersed Ruthenium in Mesoporous Crystal TiO_2 as an Advanced Electrocatalyst for Hydrogen Evolution Reaction. *J. Am. Chem. Soc.* **2018**, *140*, 5719–5727. [CrossRef]
36. Xiao, Y.X.; Ying, J.; Tian, G.; Yang, X.; Zhang, Y.X.; Chen, J.B.; Wang, Y.; Symes, M.D.; Ozoemena, K.I.; Wu, J.; et al. Hierarchically Fractal PtPdCu Sponges and their Directed Mass- and Electron-Transfer Effects. *Nano Lett.* **2021**, *21*, 7870–7878. [CrossRef]
37. Shen, L.; Ying, J.; Ozoemena, K.I.; Janiak, C.; Yang, X.-Y. Confinement Effects in Individual Carbon Encapsulated Nonprecious Metal-Based Electrocatalysts. *Adv. Funct. Mater.* **2022**, *32*, 2110851. [CrossRef]
38. Zhao, Z.; Liu, X.; Li, S.; Mao, Y.; Feng, Z.; Ke, W.; Liu, F. Study on Strengthening and Toughening of Mechanical Properties of Mg-Li Alloy by Adding Non-Rare-Earth Elements Al and Si. *JOM* **2022**, *74*, 2554–2565. [CrossRef]
39. Bai, C.; Xiang, J.; Zhao, Z.; Luo, S.; Li, L. Improvement of hydrogen production performance by in situ doping of carbon nanotubes into TiO_2 materials. *J. Nanopart. Res.* **2022**, *24*, 61. [CrossRef]
40. Yang, F.; Wen, L.Y.; Peng, Q.; Zhao, Y.; Xu, J.; Hu, M.L.; Zhang, S.F.; Yang, Z.Q. Prediction of structural and electronic properties of Cl-2 adsorbed on TiO_2(100) surface with C or CO in fluidized chlorination process: A first-principles study. *J. Cent. South Univ.* **2021**, *28*, 29–38. [CrossRef]
41. Nie, X.J.; Yin, S.Q.; Duan, W.C.; Zhao, Z.L.; Li, L.; Zhang, Z.Q. Recent Progress in Anodic Oxidation of TiO_2 Nanotubes and Enhanced Photocatalytic Performance: A Short Review. *Nano* **2021**, *16*, 2130002. [CrossRef]
42. Zhang, R.; Wu, H.; Lin, D.D.; Pan, W. Preparation of Necklace-Structured TiO_2/SnO_2 Hybrid Nanofibers and Their Photocatalytic Activity. *J. Am. Ceram. Soc.* **2009**, *92*, 2463–2466. [CrossRef]
43. Zhao, J.; Zhang, J.L.; Wang, L.; Lyu, S.S.; Ye, W.L.; Xu, B.B.; Qiu, H.; Chen, L.X.; Gu, J.W. Fabrication and investigation on ternary heterogeneous MWCNT@TiO_2-C fillers and their silicone rubber wave-absorbing composites. *Compos. Pt. A-Appl. Sci. Manuf.* **2020**, *129*, 8. [CrossRef]
44. Hu, G.; Xiao, Y.; Ying, J. Nano-SiO2 and Silane Coupling Agent Co-Decorated Graphene Oxides with Enhanced Anti-Corrosion Performance of Epoxy Composite Coatings. *Int. J. Mol. Sci.* **2021**, *22*, 11087. [CrossRef]
45. Zheng, J.Y.; Lyu, Y.H.; Wang, R.L.; Xie, C.; Zhou, H.J.; Jiang, S.P.; Wang, S.Y. Crystalline TiO_2 protective layer with graded oxygen defects for efficient and stable silicon-based photocathode. *Nat. Commun.* **2018**, *9*, 3572. [CrossRef]
46. Zeng, L.; Cui, H.; Peng, H.; Sun, X.; Liu, Y.; Huang, J.; Lin, X.; Guo, H.; Li, W.-H. Oleophobic interaction mediated slippery organogels with ameliorated mechanical performance and satisfactory fouling-resistance. *J. Mater. Sci. Technol.* **2022**, *121*, 227–235. [CrossRef]
47. Liu, Y.; Cui, H.; Wang, X.; Ouyang, G.; Guo, H. Hydrogels with Reversible Heat-Trained Toughness for Convenient Manufacture. *Adv. Mater. Technol.* **2022**, *110*, 2201346. [CrossRef]
48. Zeng, L.; Lin, X.; Li, P.; Liu, F.-Q.; Guo, H.; Li, W.-H. Recent advances of organogels: From fabrications and functions to applications. *Prog. Org. Coat.* **2021**, *159*, 106417. [CrossRef]
49. Zhou, W.; Sun, F.; Pan, K.; Tian, G.; Jiang, B.; Ren, Z.; Tian, C.; Fu, H. Well-Ordered Large-Pore Mesoporous Anatase TiO_2 with Remarkably High Thermal Stability and Improved Crystallinity: Preparation, Characterization, and Photocatalytic Performance. *Adv. Funct. Mater.* **2011**, *21*, 1922–1930. [CrossRef]

coatings

Article

Cellulose-Graphene Bifunctional Paper Conservation Materials: For Reinforcement and UV Aging Protection

Peng Tian [1,*], Meirong Shi [2], Jingmin Hou [2] and Peng Fu [2,*]

[1] School of Art and Media, Xi'an Technological University, Xi'an 710021, China
[2] Shaanxi Institute for the Preservation of Culture Heritage, Xi'an 710075, China
* Correspondence: tpmax1982@126.com (P.T.); fupeng@snnu.edu.cn (P.F.)

Abstract: Paper artifacts have unique cultural and historical values. However, over time, many paper artifacts appear with disease characteristics such as embrittlement and photoaging, losing the most fundamental function of the literature archive. The reinforcement handling of degraded paper artifacts is, therefore, a necessary measure to extend their service life, the key to which lies in the reinforcement and prevention of photoaging. This paper intended to use graphene oxide (GO) as a UV protective agent, carboxymethyl cellulose (CMC) as a reinforcement, and polyethyleneimine (PEI) as a modifier. In this work, the amino-modified graphene oxide carboxymethyl cellulose composite (CMC-aGO) was prepared by chemical modification, which was used as bifunctional paper protection material with anti-ultraviolet and reinforcement. It showed excellent performance in both tensile strength testing and UV resistance testing. The CMC-aGO raw material is low cost, colorless, transparent, simple to synthesize, convenient to operate, and is an excellent conservation material with dual functions of UV aging protection and paper reinforcement.

Keywords: graphene oxide; carboxymethyl cellulose; UV aging protection; paper reinforcement

Citation: Tian, P.; Shi, M.; Hou, J.; Fu, P. Cellulose-Graphene Bifunctional Paper Conservation Materials: For Reinforcement and UV Aging Protection. *Coatings* **2023**, *13*, 443. https://doi.org/10.3390/coatings13020443

Academic Editor: Csaba Balázsi

Received: 11 January 2023
Revised: 9 February 2023
Accepted: 14 February 2023
Published: 15 February 2023

1. Introduction

Paper cultural relics, in the form of archives, book drawings, newspapers, literature, etc., are commonly used to record valid information. As one of the important carriers of culture, science, politics, education, and economy, paper cultural relics have an immeasurable value of their own existence [1–4]. However, over time, some paper cultural relics due to natural aging, acidification, ultraviolet, and other internal and external factors lead to diseases such as yellowing, crispness, and mildew, which endanger the potential value of paper itself. The main components of paper are cellulose, lignin, and hemicellulose, as well as additives (starch, minerals, synthetic polymers, etc.) [5–9]. The aging factors include internal and external factors. Internal factors include the hydrolysis and oxidation of cellulose, which decreases the van der Waals force and hydrogen bond force between cellulose molecules, resulting in decreased intermolecular polymerization and paper mechanical strength. Extrinsic factors of aging in paper artifacts include temperature, humidity, sunlight, air pollution, microbial influences, etc. [2]. Among them, light is extremely hazardous to paper and literal pigments. UV light in daylight exposure is highly destructive. It has the characteristics of long wavelength and high energy, which may cause the change in high energy state and the transfer of energy between the paper fiber and pigment molecules, making it undergo a series of photochemical reactions [9–12]. This caused the paper fiber structure to oxidatively degrade, making the paper polymerization degree overall decline, and making its mechanical properties less stable, eventually causing irreversible damage to the paper. Therefore, developing reinforcement materials with UV aging resistance is of great significance to protect paper-based artifacts [13–15].

In the field of paper-based cultural relics protection, organic polymer materials have become a research hotspot due to their excellent reinforcement effect, simple operation, and rapid effect, mainly including synthetic polymer materials and natural polymer materials. Compared with synthetic polymer materials, natural polymer materials stand out for their advantages in safety, environmental friendliness, compatibility with paper, and reversibility, and become a hotspot of competitive research among researchers for the protection of paper artifacts. Cellulose is the most abundant and inexhaustible renewable resource in nature. It has the advantages of low cost, high quality, environmental protection, and safety [16,17]. In addition, cellulose itself is the main component of paper, and its physical and chemical properties are similar. When external environment conditions change, there is no stress between cellulose and paper fiber materials, which causes secondary damage to the paper. Basically, to meet the requirements of paper cultural relics protection for reinforcement materials, the principle of "repair the old as the old, keep the original appearance" should be followed and may also have the characteristics of aging resistance, reversibility, and paper compatibility.

Among many paper reinforcement materials, carboxymethylcellulose (CMC), as a water-soluble cellulose derivative, has excellent water solubility, viscosity, diffusibility, stability, and environmental friendliness [18]. It can be used for paper reinforcement and is widely used in this field. CMC molecules are intertwined due to their special long-chain structure [19]. The film formed by CMC has excellent toughness, flexibility, and transparency and can block oil and water. In addition, this weak alkaline solution can neutralize acidic substances in the paper and slow the progress of paper acidification and aging. Meanwhile, a large number of functional groups in the CMC structure are able to interact with hydroxyl or hydrogen bonds in the cellulose structure of the paper and are able to form a stable fibrous network on the surface of the paper, thereby playing a strengthening role [20,21]. Chen et al. found that the carboxymethyl cellulose-modified material was used to simulate the protection of paper-based cultural relics, and the synthetic carboxymethyl cellulose fluoride acetate acrylic copolymer lotion was coated on the paper. The research found that the tensile strength of the paper was increased 4.61 times, the water resistance of the paper was also greatly improved, and the reinforcing material had reversibility and certain anti-aging properties [22].

In response to the effect of UV light, graphene has excellent UV protection performance and has been applied in artifact protection in recent years. Graphene is a single-atom layer crystal constructed by hybridizing carbon atoms in sp2 [23–25]. Because of the transition effect of electrons located in the π state, graphene can absorb UV light in the range of 100~281 nm, exhibit an absorption peak at 250 nm located in the UV light region, and reflect UV light with a wavelength greater than 281 nm [26–28]. However, when graphene oxide is used directly to modify materials because the higher surface energy and van der Waals forces make it less dispersible in the matrix and easy to agglomerate, this greatly reduces the modification effect. Numerous studies have shown that the amino functionalization modification of graphene oxide can improve the dispersity of graphene oxide materials. Meanwhile, more reactive sites are added, which is helpful for strengthening interactions with the polymer matrix. Therefore, it may be able to take advantage of the preventive protection with paper artifacts, thereby weakening the damage of the paper by UV light [29–32].

Based on this, this work utilized aminolated graphene oxide (aGO) as an anti-UV aging agent and carboxymethyl cellulose as a paper reinforcement. The related research of carboxymethylcellulose-amino graphene- composite (CMC-aGO) was expected to provide a theoretical basis for the development of anti-light aging and reinforcement materials for paper relics. The innovations of this work are as follows: (1) Surface modification of CMC with PEI enriches the surface with amino groups and provides an anchoring point for graphene nanosheets; (2) The composite material of graphene and CMC is prepared by simple technology, and the raw material is cheap and easy to obtain, which has practicability and application prospect; (3) The development of reinforcement materials with UV resistance properties enables restorative materials to function as prophylactic protection on the basis of the reinforcement of damaged paper.

2. Materials and Methods

2.1. Chemicals

CMC, epichlorohydrin, PEI (molecular weight about 600 g mol^{-1}), and KBr were purchased from Shanghai Aladdin Biochemical Technology Co., Ltd. (Shanghai, China). The reagents were all analytical and pure, and the water used for the experiment was deionized water. The simulated pigments (vermilion, malachite, ultramarine, and gamboge), alum gelatin, and rice paper were purchased from Beijing Tianya Pigment Co., Ltd. (Beijing, China).

2.2. Preparation of Samples

2.2.1. Synthesis of CMC-aGO

First, 10 mg of GO was added into 140 mL of deionized water, which was pulverized by a homogenizer for 60 min, and ultrasonic dispersion for 30 min to obtain go dispersion with a mass fraction of 6.66%, which was then placed at 80 °C for heating. Then, 0.4 g PEI and 0.015 g KOH were added to the above GO dispersion to obtain the GO-PEI mixture by continuous heating and stirring at 80 °C for 12 h. Next, 0.1 g CMC was added to the above GO-PEI mixture with stirring for 1 h; then 0.5 g epichlorohydrin was added dropwise to the above reaction system to react at 80 °C for 12 h (epichlorohydrin acted as a crosslinking agent [33]). Finally, they were centrifuged, and the obtained solid was washed with ethanol, deionized water repeatedly, and freeze-dried for 48 h. The CMC-GO is directly mixed with CMC and GO, and its proportion is the same as that of CMC-aGO.

2.2.2. Preparation of Simulated Samples

The rice paper was trimmed to the size of 15~240 mm, and the diluted pigments were, respectively, painted evenly on the paper samples, thoroughly shade dried, and protected from light. An amount of 5 mL of the formulated CMC and CMC-aGO dispersions was placed in the spray gun, sprayed evenly on the rice paper with pigments and those without pigments, and placed on the yarn mesh to air dry naturally. These samples were used as simulated samples for the color difference test, tensile strength, and light resistance tests.

2.3. Methods of Performance Testing

2.3.1. Instrument Parameter

The molecular structures of CMC, aGO, and CMC-aGO were characterized by Fourier transform infrared spectroscopy. (FTIR, Vertex70, Bruker, Karlsruhe, Germany). Spectral range: 500–4000 cm^{-1}. Scanning electron microscopy (SEM, Hitachi High-tech Co. Ltd., SU-8020, Tokyo, Japan) was used to study the microstructure of the samples. An X-Rite VS450 non-contact spectrophotometer was used to test the color difference of the paper samples.

2.3.2. Dry-Heat Accelerated Aging Experiments

According to GB/T464-2008 (dry heat accelerated aging of paper and cardboard), the paper samples were hung in blueparbho-401, a dry heat aging box, which was set to 105 °C, and the aging time was 3 days.

2.3.3. Wet-Heat Accelerated Aging Experiments

According to GB/T22894-2008 (wet heat treatment of accelerated aging of paper and cardboard at 80 °C and 65% Relative Humidity), the paper samples were hung up in an HCP wet heat aging box for accelerated aging experiments, which was set at 80 °C and 65% relative humidity for 3 days.

2.3.4. UV Aging

The treated samples were placed in a UV aging oven under the following conditions: 40 W power, 313 nm wavelength, and a distance of 50 mm between the surface and the plane of the UV lamp. They were, respectively, irradiated for 3 days, 6 days, and 9 days and stored in the dark for 24 h for the performance test of paper sample tensile strength testing and color difference.

2.3.5. Tensile Strength Testing

Tensile strength testing was used to measure the ability of a paper to resist stretching from external forces. In accordance with GB/T12914-2008 (tensile strength determination of paper and sheets), the measured sheets were cut to size 15 mm × 240 mm. A QT-1136PC universal material testing machine (Xie Qiang Instrument Manufacturing Co. Ltd., Shanghai, China) was used with a tensile rate of 20 mm/min, and the paper was stretched to fracture so that the tensile strength S (kN/m) was obtained. Tensile strength tests were performed along the transverse and longitudinal sheets, respectively.

2.3.6. Color Difference Testing

A CIE L × a × b color coordinate system was used to characterize the color changes before and after paper sample treatment and accelerated aging. The smaller the chromatic difference value (ΔE) of the paper represents, the smaller the color change before and after sample reinforcement and accelerated aging. The CIE L × a × b color system chromatic difference value is calculated as follows:

$$\Delta E = [(\Delta L)^2 + (\Delta a)^2 + (\Delta b)^2]^{1/2}$$

Among them, ΔL represents the difference in light acuity, Δa represents the red green deviation, and Δb represents the yellow blue deviation [34].

3. Results

3.1. Characterization of CMC-aGO

The present work was intended to proceed through the synthesis of CMC-aGO (Figure 1). On the one hand, carboxymethyl cellulose cross-links with the inner fiber of the paper, which in turn achieves the effect of fill reinforcement. On the other hand, the polyethyleneimine (PEI)-modified graphene oxide surface contains a high density of amino groups, which easily form bonds with the cellulose surface carboxymethyl, such that GO nanosheets adhere to the cellulose surface; the introduction of graphene can achieve the effect of protecting against UV light.

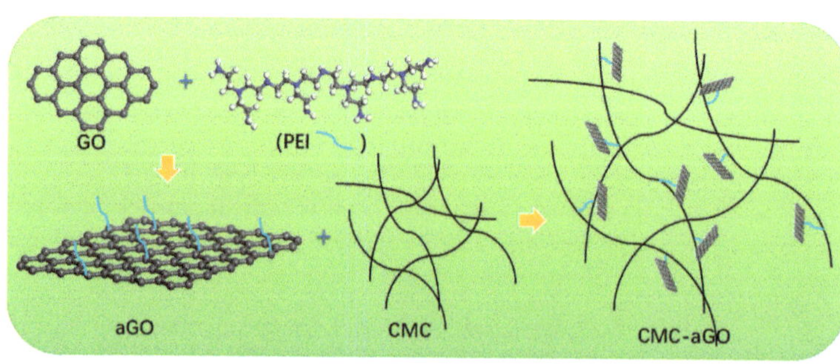

Figure 1. Schematic of the synthetic procedure for CMC-aGO.

The microtopography of CMC-aGO was observed by SEM. As shown in Figure 2a, CMC-aGO consisted of a large number of nanofibers with diameters of ~10 nm. The nanofibers were randomly arranged and overlapped to form a coherent three-dimensional porous network structure. The GO layers are closely bonded to a large number of BC nanofibers without agglomeration (as shown in the color block diagrams of Figure 2a). In terms of macroscopic morphology, CMC-aGO was uniformly dispersed after 12 h rest.

However, the CMC-GO, which directly mixes GO and CMC, was obviously separated (Figure 2b).

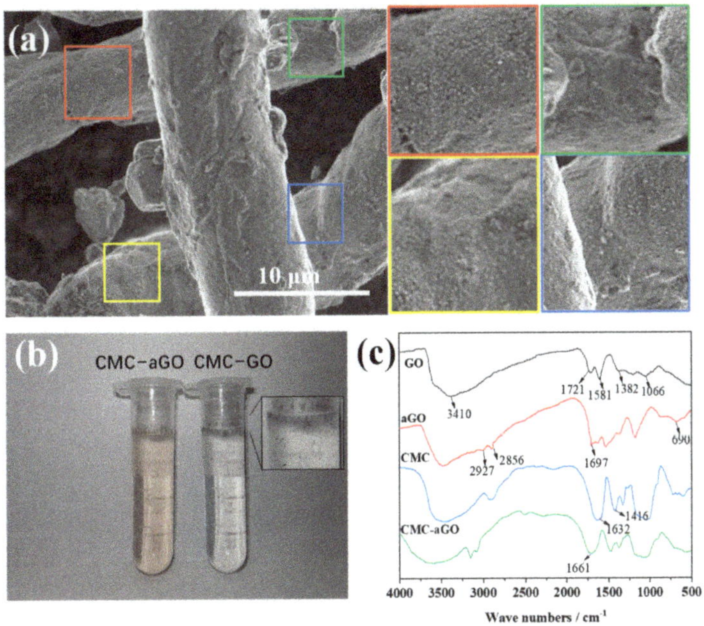

Figure 2. (**a**) SEM of CMC-aGO (Colored blocks on the right are local magnifications); (**b**) Photo of CMC-aGO and CMC-GO; (**c**) FT-IR of GO, aGO, CMC, and CMC-aGO.

Then, in order to further verify the interaction between graphene, PEI, and cellulose, a series of materials were tested by FT-IR. Figure 2c shows the FT-IR diagrams of GO, aGO, CMC, and CMC-aGO. The absorption peaks of GO at 1066, 1381, 1581, and 1721 cm^{-1} were, respectively, generated by C–O, C-OH, C=C, and C-O-C vibration. The wider absorption peak at 3410 cm^{-1} is the expansion vibration of -OH. The absorption peaks of PEI-modified GO at 2856 and 2927 cm^{-1} were caused by the C-H tensile vibration of aliphatic and aromatic groups, and the characteristic peaks at 1697 and 669 cm^{-1} were caused by N-H bending and swinging, respectively. This was caused by the amidation reaction between amino groups in PEI and carboxyl groups on the GO surface, which indicates that PEI was modified on the GO surface. The peaks of CMC at 1632 cm^{-1} and 1416 cm^{-1} are asymmetric and symmetric expansion vibration absorption peaks of -COOH, respectively. The FT-IR spectra of CMC-aGO show that the N-H characteristic peak disappeared and the characteristic peak of -COOH decreased. The absorption peak at 1661 cm^{-1} may be due to the amidation of -CONH- between -NH_2 in PEI and -COOH in CMC or GO, which indicates that GO and PEI successfully grafted onto the CMC surface [35,36].

The micromorphology of the paper cultural relics is shown in Figure 3a, the paper fibers are intertwined, and there are a large number of voids among the fibers. Excellent reinforcing materials can penetrate into the voids between the fibers and cross-link with the internal fibers of the paper to achieve the effect of filling and reinforcing. There were obvious deterioration signs, such as the breakage and fracture of the paper fiber before reinforcement, and a large number of voids exist among the fibers (Figure 3a, red dotted circle marker in Figure 3a). After CMC-aGO reinforcement, the CMC fibers formed cross-linking structures between the paper fibers, which reduced the porosity of the paper samples and reinforced the fragile paper (Figure 3b). SEM showed that the size of paper fibers did not change significantly before and after CMC repair, and the surface of the repaired sample was smooth.

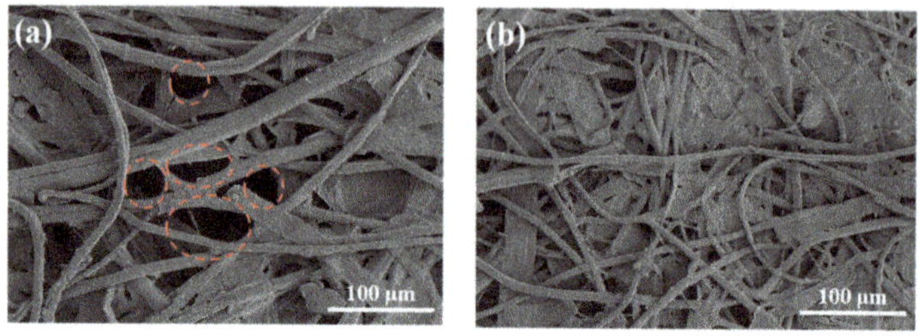

Figure 3. SEM of (**a**) Paper fibers; (**b**) Paper fiber reinforcement by CMC-aGO. (red circles represent interfibrillar voids).

3.2. Performance Characterization of CMC-aGO

The color change is an important indicator to evaluate the suitability of reinforcement materials for paper reinforcement. The color difference measurements of the paper and pigments before and after reinforcement were tested. As shown in Figure 4a, the color difference of the paper changed after being treated with CMC and CMC-aGO. After CMC and CMC-aGO treatment, the paper color difference values were 1.3 and 1.5, respectively, and the color difference standard ($\Delta E = 3$) was less than three, which belonged to the range of color difference tolerance. It shows that CMC had no significant effect on the original appearance of the paper. Then, the color difference effects of the four pigments under the reinforcement of CMC and CMC-aGO were tested. As shown in Figure 4b, the influence of CMC and CMC-aGO on the color difference of vermilion, malachite, ultramarine, and gamboge pigments were studied. After the minor addition of graphene was added, the ΔE of vermilion and malachite samples treated by CMC-aGO were 2.3 and 2.2, respectively, which is slightly higher than that of the samples treated by CMC. However, an ΔE less than three belonged to the range of color difference tolerance. The color difference between CMC-aGO on ultramarine and gamboge was less than that of CMC, and the effect of pure color was obvious. The results showed that the addition of a small amount of graphene did not cause any change in the color difference between the paper and pigment.

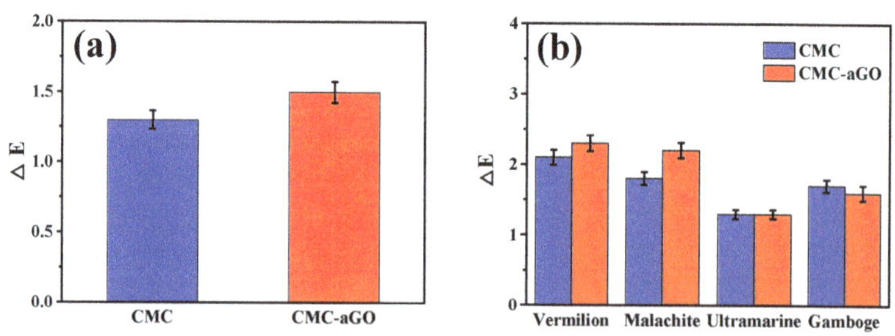

Figure 4. The color difference of (**a**) The paper and (**b**) Pigments before and after reinforcement.

The tensile strength of paper is one of the important indexes to evaluate the physical performance of the paper. CMC and CMC-aGO were used to reinforce the rice paper, while the alum gelatin was used as a control. The tensile strength test results of the paper after reinforcement are shown in Figure 5; the tensile strength of the paper after reinforcement with alum gelatin, CMC and CMC-aGO was significantly larger compared to that of the untreated rice paper. Among them, CMC-aGO showed the most excellent performance with a tensile strength of 2.2 kN/m and 2.9 kN/m in the cross and machine direction, which were significantly better than alum gelatin (1.4 kN/m, 2.3 kN/m) and CMC (1.7 kN/m, 2.6 kN/m).

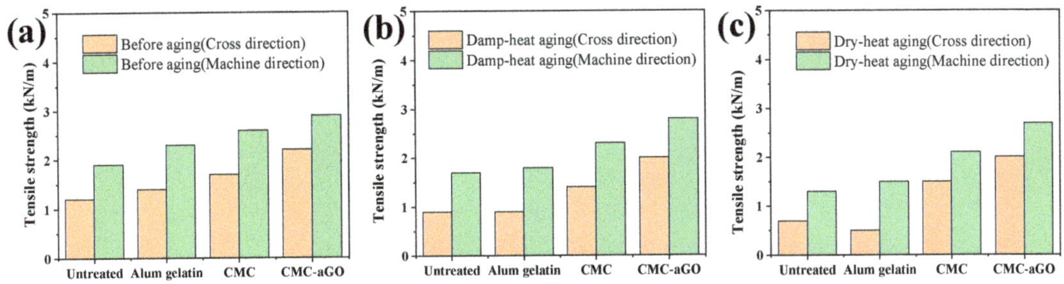

Figure 5. Tensile Strength of paper (**a**) Before aging, (**b**) After damp-heat aging, and (**c**) Dry-heat aging.

Then, dry heat aging and wet heat aging treatments were performed on the paper treated with different reinforcement agents. The results showed that, compared with the blank samples, CMC-aGO-treated samples were least affected by dry-heat aging and wet-heat aging, and the samples treated with alum had the greatest influence. After dry heat aging, the cross-direction strength and machine direction of CMC-aGO treated samples decreased by 9% and 6.9%, respectively, and after wet-heat aging, the cross and machine direction strength decreased by 9% and 3.4%, respectively. While significantly better than CMC, this may have been due to the interactions between the aGO and CMC with the paper fiber, the oxygen-containing groups, and hydrogen bonds in the CMC aGO interacting with the paper cellulose, thus enabling the formation of the stable fibrous network on the surface of the paper, and achieving the filling reinforcement at the damaged paper fiber pores and fracture sites, increasing the strength and toughness of the fibers, and reinforcement of the paper.

The reinforcement mechanism was analyzed. The paper fiber strength and the binding strength between the fibers are the main factors that affect the paper's mechanical strength. Fiber length and fiber interlacing factors also contribute to the paper's mechanical strength. These factors affect the strength performance index of the paper in a complex way, such as the tensile strength, the number of folds, and tear strength. Usually, the binding strength between cellulose is the most dominant influencing factor. After CMC treatment, the inter-fiber bonding strength is strengthened.

Some pigments are prone to photo-oxidation reaction under UV light, which excites the double bond of part of the excited state to react with the oxygen in the air, leading to a change in pigment color, a faded appearance, and a discoloration phenomenon. To explore the UV light-resistant properties of reinforcement materials, a series of tests were carried out. It is known from Figure 6a that the pigment color difference value ΔE of each group remained largely unchanged after 90 days of UV aging for CMC-aGO treated samples compared to CMC-treated samples during the UV aging test. Among them, CMC-aGO showed the best color fixation effect on vermilion and gamboge. The CMC-treated samples showed more variable color difference values starting from 30 days, all of which exceeded the values of ΔE = 3, especially for the vermilion pigment. This phenomenon might be attributed to the lack of UV resistance properties in CMC materials, and the vermilion was prone to crystal phase transition to beta-mercury sulfide (β-Hg S) under UV light conditions. Ultraviolet light has higher energy than is required for crystal phase transformation, which can make the cinnabar crystal phase change rapidly and color become darker. Graphene in CMC-aGO absorbs some ultraviolet rays, effectively delaying the damage caused by ultraviolet rays.

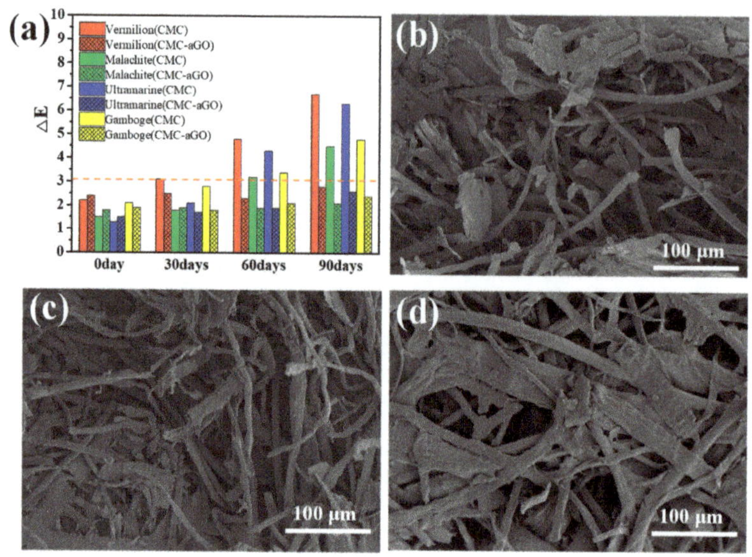

Figure 6. (**a**) Color difference of four pigments treated with CMC and CMC-aGO after UV aging. SEM of (**b**) The blank group, (**c**) CMC, and (**d**) CMC-aGO-treated paper after UV aging.

The microstructure test results show that (Figure 6b–d) some fiber structure breakage occurred after aging, among which, the damaged fibers of the blank group (Figure 6b) and CMC (Figure 6c) were more serious after aging. While the CMC-aGO-treated samples largely kept their fibrous structures intact after UV aging, only a few fiber fracture phenomena appeared. This phenomenon further confirmed the anti-UV effect of GO in CMC-aGO.

4. Conclusions

In summary, CMC with aminolated graphene oxide composites was successfully synthesized by a simple synthetic method, which was applied with paper reinforcement and protection against UV light. The results show that the PEI-modified aGO surface contains high-density amino groups, which are easy to bond with carboxymethyl groups on the surface of cellulose. Graphene nano-sheets adhere to the surface of cellulose and effectively improve the dispersion of graphene in the water. CMC in the composite materials crosslinks with the fibers inside the paper to achieve filling and reinforcement effects, and GO in the composite materials has an anti-ultraviolet effect. By testing the color difference in paper reinforcement and pigments, it is shown that the appearance of paper cultural relics was not affected by the small amount of graphene added. The reinforcement and ultraviolet resistance of CMC-aGO were verified by tensile tests and simulated aging tests. The development of CMC-aGO has great application potential in the protection of cultural heritage and is of great significance to the development of paper conservation materials.

Author Contributions: Conceptualization, P.T.; Methodology, M.S.; Investigation, J.H.; Writing—original draft, P.F. All authors have read and agreed to the published version of the manuscript.

Funding: This research received no external funding.

Institutional Review Board Statement: Not applicable.

Informed Consent Statement: Not applicable.

Data Availability Statement: Not applicable.

Conflicts of Interest: The authors declare no conflict of interest.

References

1. He, B.; Lin, Q.X.; Chang, M.M.; Liu, C.F.; Fan, H.M.; Ren, J.L. A new and highly efficient conservation treatment for deacidification and strengthening of aging paper by in-situ quaternization. *Carbohydr. Polym.* **2019**, *209*, 250–257. [CrossRef] [PubMed]
2. Zervos, S.; Alexopoulou, I. Paper conservation methods: A literature review. *Cellulose* **2015**, *22*, 2859–2897. [CrossRef]
3. Barnat-Hunek, D.; Grzegorczyk-Franczak, M.; Klimek, B.; Pavlikova, M.; Pavlik, Z. Properties of multi-layer renders with fly ash and boiler slag admixtures for salt-laden masonry. *Constr. Build. Mater.* **2021**, *278*, 122366. [CrossRef]
4. Noake, E.; Lau, D.; Nel, P. Identification of cellulose nitrate based adhesive repairs in archaeological pottery of the university of melbourne's middle eastern archaeological pottery collection using portable ftir-atr spectroscopy and pca. *Herit. Sci.* **2017**, *5*, 3. [CrossRef]
5. Akyuz, T.; Akyuz, S.; Balci, K.; Gulec, A. Investigations of historical textiles from the imperial pavilion (hunkar kasri) of the new mosque eminonu-istanbul (turkey) by multiple analytical techniques. *J. Cult. Herit.* **2017**, *25*, 180–184. [CrossRef]
6. Bertolini, L.; Carsana, M.; Redaelli, E. Conservation of historical reinforced concrete structures damaged by carbonation induced corrosion by means of electrochemical realkalisation. *J. Cult. Herit.* **2008**, *9*, 376–385. [CrossRef]
7. Gorassini, A.; Adami, G.; Calvini, P.; Giacomello, A. Atr-ftir characterization of old pressure sensitive adhesive tapes in historic papers. *J. Cult. Herit.* **2016**, *21*, 775–785. [CrossRef]
8. Havlinova, B.; Katuscak, S.; Petrovicova, M.; Makova, A.; Brezova, V. A study of mechanical properties of papers exposed to various methods of accelerated ageing. Part I: The effect of heat and humidity on original wood-pulp papers. *J. Cult. Herit.* **2009**, *10*, 222–231. [CrossRef]
9. Yang, Z.H.; Zhang, J.J.; Qi, Y.P.; Shen, Y.F.; Li, H. Application of hydrophobically modified hydroxyethyl cellulose-methyl methacrylate copolymer emulsion in paper protection. *Nord. Pulp. Pap. Res. J.* **2022**, *37*, 282–289. [CrossRef]
10. Wu, X.; Mou, H.Y.; Fan, H.M.; Yin, J.Y.; Liu, Y.B.; Liu, J.A. Improving the flexibility and durability of aged paper with bacterial cellulose. *Mater. Today Commun.* **2022**, *32*, 103827. [CrossRef]
11. Shehata, M.; Navarra, M.; Klement, T.; Lachemi, M.; Schell, H. Use of wet cellulose to cure shotcrete repairs on bridge soffits. Part 1: Field trial and observations. *Can. J. Civ. Eng.* **2006**, *33*, 807–814. [CrossRef]
12. Luo, Y.B. Durability of chinese repair bamboo papers under artificial aging conditions. *Stud. Conserv.* **2019**, *64*, 448–455. [CrossRef]
13. Henniges, U.; Angelova, L.; Schwoll, S.; Smith, H.; Bruckle, I. Microfibrillated cellulose films for mending translucent paper: An assessment of film preparation and treatment application options. *J. Inst. Conserv.* **2022**, *45*, 36–51. [CrossRef]
14. Fu, L.N.; Zhang, J.; Yang, G. Present status and applications of bacterial cellulose-based materials for skin tissue repair. *Carbohydr. Polym.* **2013**, *92*, 1432–1442. [CrossRef]
15. Egertsdotter, E.M.U.; Aidun, C.K. Renewable materials for tissue repair/biocompatibility of cellulose nanocrystals. In Proceedings of the 5th IASTED International Conference on Biomedical Engineering, Innsbruck, Austria, 14–16 February 2007.
16. Dadabayev, R.; Malka, D. A visible light rgb wavelength demultiplexer based on polycarbonate multicore polymer optical fiber. *Opt. Laser Technol.* **2019**, *116*, 239–245. [CrossRef]
17. Gelkop, B.; Aichnboim, L.; Malka, D. Rgb wavelength multiplexer based on polycarbonate multicore polymer optical fiber. *Opt. Fiber Technol.* **2021**, *61*, 102441. [CrossRef]
18. Borges, I.D.; Casimiro, M.H.; Macedo, M.F.; Sequeira, S.O. Adhesives used in paper conservation: Chemical stability and fungal bioreceptivity. *J. Cult. Herit.* **2018**, *34*, 53–60. [CrossRef]
19. Rahman, M.S.; Hasan, M.S.; Nitai, A.S.; Nam, S.; Karmakar, A.K.; Ahsan, M.S.; Shiddiky, M.J.; Ahmed, M.B.J.P. Recent developments of carboxymethyl cellulose. *Polymers* **2021**, *13*, 1345. [CrossRef]
20. Sanchez, E.; Sanz, V.; Bou, E.; Monfort, E. Carboxymethylcellulose used in ceramic glazes part iii: Influence of cmc characteristics on glaze slip and consolidated glaze layer properties. *Cfi-Ceram. Forum Int.* **1999**, *76*, 24–27.
21. Petrea, P.; Ciolacu, F.; Ciovica, S. The evaluation of some consolidation agents applied in the conservation of graphical documents. *Eur. J. Sci. Theol.* **2010**, *6*, 67–75.
22. Chen, Q.; Wen, W.-Y.; Qiu, F.-X.; Xu, J.-C.; Yu, H.-Q.; Chen, M.-L.; Yang, D.-Y. Preparation and application of modified carboxymethyl cellulose si/polyacrylate protective coating material for paper relics. *J. Chem. Pap.* **2016**, *70*, 946–959. [CrossRef]
23. Farivar, F.; Yap, P.L.; Tung, T.T.; Losic, D. Highly water dispersible functionalized graphene by thermal thiolene click chemistry. *Materials* **2021**, *14*, 2830. [CrossRef] [PubMed]
24. Costa, M.C.F.; Marangoni, V.S.; Ng, P.R.; Nguyen, H.T.L.; Carvalho, A.; Neto, A.H.C. Accelerated synthesis of graphene oxide from graphene. *Nanomaterials* **2021**, *11*, 551. [CrossRef] [PubMed]
25. Binder, J.; Rogoza, J.; Tkachenko, L.; Pasternak, I.; Sitek, J.; Strupinski, W.; Zdrojek, M.; Baranowski, J.M.; Stepniewski, R.; Wysmolek, A. Suspended graphene on germanium: Selective local etching via laser-induced photocorrosion of germanium. *2D Mater.* **2021**, *8*, 035043. [CrossRef]
26. Rochford, C.; Kumar, N.; Liu, J.W.; Zhao, H.; Wu, J. All-optical technique to correlate defect structure and carrier transport in transferred graphene films. *ACS Appl. Mater. Inter.* **2013**, *5*, 7176–7180. [CrossRef]
27. Duong, D.L.; Han, G.H.; Lee, S.M.; Gunes, F.; Kim, E.S.; Kim, S.T.; Kim, H.; Ta, Q.H.; So, K.P.; Yoon, S.J.; et al. Probing graphene grain boundaries with optical microscopy. *Nature* **2012**, *490*, 235–239. [CrossRef]
28. Chen, Z.Y.; Zhang, Y.H.; Zhang, H.R.; Sui, Y.P.; Zhang, Y.Q.; Yu, G.H.; Jin, Z.; Liu, X.Y. Probing graphene grain boundaries and enhancing the electrical properties of graphene films by pt modification. *Mater. Lett.* **2014**, *131*, 53–56. [CrossRef]

29. Saladino, M.L.; Markowska, M.; Carmone, C.; Cancemi, P.; Alduina, R.; Presentato, A.; Scaffaro, R.; Bialy, D.; Hasiak, M.; Hreniak, D.; et al. Graphene oxide carboxymethylcellulose nanocomposite for dressing materials. *Materials* **2020**, *13*, 1980. [CrossRef]
30. Ponnuchamy, V.; Esakkimuthu, E.S. Density functional theory study of lignin, carboxymethylcellulose and unsustainable binders with graphene for electrodes in lithium-ion batteries. *Appl. Surf. Sci.* **2022**, *573*, 151461. [CrossRef]
31. Karimzadeh, Z.; Javanbakht, S.; Namazi, H. Carboxymethylcellulose/mof-5/graphene oxide bio-nanocomposite as antibacterial drug nanocarrier agent. *Bioimpacts* **2019**, *9*, 5–13. [CrossRef]
32. Araque, E.; Villalonga, R.; Gamella, M.; Martinez-Ruiz, P.; Sanchez, A.; Garcia-Baonza, V.; Pingarron, J.M. Water-soluble reduced graphene oxide-carboxymethylcellulose hybrid nanomaterial for electrochemical biosensor design. *ChemPlusChem* **2014**, *79*, 1334–1341. [CrossRef]
33. Guan, H.; Li, J.; Zhang, B.; Yu, X. Synthesis, properties, and humidity resistance enhancement of biodegradable cellulose-containing superabsorbent polymer. *J. Polym.* **2017**, *2017*, 3134681. [CrossRef]
34. Bourguignon, E.; Tomasin, P.; Detalle, V.; Vallet, J.-M.; Labouré, M.; Olteanu, I.; Favaro, M.; Chiurato, M.A.; Bernardi, A.; Becherini, F. Calcium alkoxides as alternative consolidants for wall paintings: Evaluation of their performance in laboratory and on site, on model and original samples, in comparison to conventional products. *J. Cult. Herit.* **2018**, *29*, 54–66. [CrossRef]
35. Yang, G.; Cao, J.; Li, L.; Rana, R.K.; Zhu, J.-J. Carboxymethyl chitosan-functionalized graphene for label-free electrochemical cytosensing. *Carbon* **2013**, *51*, 124–133. [CrossRef]
36. Singh, A.; Yadagiri, G.; Negi, M.; Kushwaha, A.K.; Singh, O.P.; Sundar, S.; Mudavath, S.L.J.I.J.o.B.M. Carboxymethyl chitosan modified lipid nanoformulations as a highly efficacious and biocompatible oral anti-leishmanial drug carrier system. *Int. J. Biol. Macromol.* **2022**, *204*, 373–385. [CrossRef]

coatings

Article

Application of Starch Based Coatings as a Sustainable Solution to Preserve and Decipher the Charred Documents

Sonali Kesarwani [1], Divya Bajpai Tripathy [1,*] and Suneet Kumar [2]

1 School of Basic and Applied Sciences, Galgotias University, Greater Noida 201312, Uttar Pradesh, India
2 Forensic Science Laboratory, Moradabad 244001, Uttar Pradesh, India
* Correspondence: divya.tripathy@galgotiasuniversity.edu.in

Abstract: Fire can be one of the most destructive elements to cause devastation. Fire can completely or partly destroy any crucial and invaluable documents, such as banknotes, books, affidavits, etc., in a couple of minutes. Moreover, the documents can also be damaged by heat, smoke, soot, and water during an accident. The burnt documents become fragile, losing their identity, which may have some evidentiary value related to the incident. Therefore, there is a strong need for processing to procure, preserve, and decipher, i.e., to restore the texts written on them. Hence, the present research focuses on developing a new method using natural polysaccharides, i.e., starch, to preserve and decipher the contents of charred documents. The most suitable concentration of starch analog was found to be 6% microwaved at 80 °C for about 10 min. As soon as the charred documents were coated with 6% starch analog, the majority of the invisible texts became visible to the naked eye in a second. Moreover, the application of a synthesized analog of polysaccharide on fragile charred documents provided an appreciable increase in strength by almost 0.1 kg/cm^2 for the coated charred documents of each paper type compared to that of non-coated ones and made them stabilized. This research also involves the use of easy and advanced handwriting recognition techniques (HCR) using an easily accessible, free platform, G-lens, that successfully recognized the majority of texts deciphered using 6% starch analog and converted them from captured images to a readable and copyable text format. Furthermore, the document visualization under VSC also gave a promising result by enhancing and deciphering the non-visible and less visible texts under flood light and white spot light at 715 and 695 long passes. Hence, this study offers an environmentally friendly, cost-effective, and sustainable approach of using a natural polysaccharide instead of synthetic polymers for the preservation and decipherment of charred documents.

Keywords: natural polysaccharide; questioned document; charred; decipherment; character recognition

Citation: Kesarwani, S.; Tripathy, D.B.; Kumar, S. Application of Starch Based Coatings as a Sustainable Solution to Preserve and Decipher the Charred Documents. *Coatings* **2023**, *13*, 1521. https://doi.org/10.3390/coatings13091521

Academic Editors: Yulan Hu and Shiqiang Fang

Received: 14 July 2023
Revised: 1 August 2023
Accepted: 25 August 2023
Published: 30 August 2023

1. Introduction

In arson cases or other fire incidents, documents, both printed and handwritten with writing instruments (generally pen), can be destroyed either accidentally or purposely due to excessive heat and smoke. During a fire incident, one or more paper placed together may burn either completely or partly due to a limited oxygen supply. The deliberately or accidentally burned documents may have an evidentiary value and may provide proof to link a crime with a criminal. Often, such documents may not completely turn into ashes due to several factors that may prevent the entire document from being burned; however, they may still make the document fragile, brittle, or black and render the writing invisible to the naked eye or unreadable, due to excessive smoke and soot from the fire [1]. Such documents need to be handled and preserved with utmost care, and the crucial content needs to be deciphered using a suitable method.

Decipherment of charred documents refers to analyzing and interpreting texts that have been damaged by fire or other heat sources. It is a specialized field of study that requires a combination of knowledge in linguistics, paleography, and chemistry. Charred documents can be found in various contexts, such as archaeological sites, historical archives and crime scenes. They may contain valuable information about ancient cultures, historical events, or personal records. However, their decipherment is often challenging because the heat can cause the ink or writing material to evaporate or fuse with the surface, making the text illegible. To decipher charred documents, experts may use a variety of techniques, such as multispectral imaging, chemical analysis, and comparative analysis with known texts. They may also rely on their knowledge of the document's language, script, and historical context. The decipherment of charred documents is important because it can reveal new insights into the past that would otherwise be lost. It can shed light on historical events, social practices, and linguistic evolution. It can also provide a deeper understanding of the people who created the documents and the cultural context in which they lived.

Before the 20th century, writing inks contained traces of metals like iron and copper as tagging agents. Therefore, for deciphering the writing written with such inks, Blagden (1787) developed a method using potassium ferrocyanide to test the nature of the ink on ancient parchment for decipherment [2]. Later, Davis (1922) [3] proposed a method using a photographic plate to decipher the content of charred documents, while Mitchell (1925) used the calcining method, a process of further burning the carbonized fragment to decipher content written with pencil or some special inks, or content that was typewritten or printed [4]. Moreover, in 1935, Mitchell used infrared (IR) light with filters and plates to enhance charred documents' content [5]. Subsequently, Radley and Grant (1940) used fluorescent oil and ultraviolet light to successfully decipher the writing on printed matter, photocopies, typescripts, and carbon copies [6]. Similarly, many other methods have been developed, such as using chloral hydrate, a 5% solution of silver nitrate, an alcohol–glycerin solution, etc., to decipher the content of charred printed and typewritten documents. Over time, the development in ink composition led to iron being removed as a constituent because of its rusting and corrosive effect on the nib of pens, hence damaging the paper on which such pens were used. Therefore, current ink compositions exclude any traces of metals and only include pigments, dyes, resins, glycerol, alcohol, oils, and fats [7]. Thus, when documents written using such inks are charred, their components like pigments, dyes, alcohol, resins, etc., burn out, leaving lubricants like oils and fats on the surface of the charred documents because of their high boiling point. This aids in the decipherment of writing by different means. In questioned document examination, the major challenge faced by forensic practitioners is the handling and stabilization of highly fragile and brittle charred documents, hence rendering the decipherment of crucial information possible.

Deciphering charred documents using thin coatings is an established technique known as "palimpsest imaging". The term "palimpsest" refers to a manuscript or piece of writing on which the original writing has been erased or obscured, and new writing has been added on top. In order to decipher charred documents, a thin coating of some specific materials such as gold or silver is applied to the surface of the document. This coating reflects light differently depending on the depth of the writing and can expose previously obscured text. The process is based on the principles of reflection of light on the document at a certain angle, which creates shadows where the text is raised. These shadows can be captured using a high-resolution camera and can then be further processed using specialized software to enhance the contrast between the text and the background. This technique has been successfully used to decipher ancient manuscripts, including the Archimedes Palimpsest, a 10th-century manuscript containing the only surviving copies of some of Archimedes' works. The technique has also been used to reveal text on charred documents from the Villa of the Papyri in Herculaneum, which was destroyed because of the eruption of Mount Vesuvius in 79 AD. However, palimpsest imaging is a powerful tool for deciphering charred documents but the exploitation of costly metals makes this procedure cost-ineffective. Moreover, this technique is limited to document decipherment and is unsuitable for their preservation.

Other chemicals used for this purpose are polyvinyl acetate (PVA), methyl methacrylate (Bed acryl) [8], acetone, alkyl-2-cyanoacrylate ester (superglue fuming) [9], ammonia solutions [10], etc. Moreover, according to Harrison, the use of PVA may hinder decipherment under infrared (IR) and ultraviolet (UV) light and also requires some preprocessing before applying PVA over charred documents [11,12]. Additional challenges associated with this approach are the chemical nature, toxic effect, low resistance to weather and moisture, poor resistance to most solvents, slow setting speed, creeping under substantial static load, etc. [11]. The application of PVA on charred documents is also a challenging task. Various researchers suggested the Pasteur pipette as a suitable and appropriate means of PVA application, while others recommended the use of a fine mister, such as a re-used perfume sprayer. In both the cases, significant caution and care need to be taken while applying the PVA in acetone solution to avoid further tempering to the charred documents due to their extremely fragile nature [13].

To overcome these challenges, researchers explored and developed an improved preservative and better procedure for its application over charred documents: natural polysaccharide coating as a green, non-toxic, cost-effective approach. This method also offers a fast and non-tedious synthetic and application technique to preserve the charred documents [14]. Moreover, transparent texture (for good visibility of texts), appropriate viscosity (for ease in application), good decipherment properties, short drying duration, and ability to provide appreciable strength to the coated charred documents were also the mandatory requisites that were found with starch-based coatings.

Polysaccharides, one of the most plentiful natural polymers, have the potential to take the place of "petroleum-based polymers", which are challenging to degrade in paper coatings. Polysaccharide molecules have a large number of hydroxyl (-OH) groups that can bind strongly with paper fibers through hydrogen bonds. Furthermore, their chemical modification can also effectively improve the mechanical barrier, thermal resistance, and hydrophobic properties of polysaccharide-based coatings and make them suitable to coat charred paper. In addition, polysaccharides can also give additional functional properties to the paper by dispersing and adhering functional fillers, e.g., conductive particles, catalytic particles, or anti-microbial chemicals, onto the coated paper surface. Starch is a polysaccharide made of two types of α-D-glucan chains: amylose and amylopectin. Starch molecules produced by each plant species have specific structures and compositions (such as the length of glucose chains or the amylose/amylopectin ratio), and the protein and fat content of the storage organs may vary significantly. The majority of natural starches are amylopectin, which has film-forming ability, but its film mechanical properties still need to be improved [15]. When used as a paper coating, pure starch still has some other drawbacks. For instance, starch is sensitive to water vapor and usually forms a brittle coating layer [16]. Pure starch also forms faults in coating layers because of residual air, which results in large surface pores. To avoid all these problems, starch is usually modified through gelatinization and etherification [17–19]. Starch is a natural biopolymer that has been extensively used in various industries such as food, pharmaceuticals, and papermaking due to its abundance, low cost, and biodegradability [20]. Moreover, it is also used in the archaeological conservation of documents and artifacts to shield them from further deterioration [21]. Starch and its derivatives are the most frequently used sizing agents [22]. Sizing is used to improve the absorption and wear characteristics of paper. The most common materials used in sizing solutions are starch, latex, polyvinyl alcohol (PVA), and carboxymethyl cellulose (CMC). Surface sizing requires modification of native starch to achieve low viscosity of the starch solution. Many kinds of modified starches, such as acid-degraded starch, oxidized starch, starch ester, cationic starch, enzyme-degraded starch, and carboxymethyl starch, are used for surface sizing [23]. Starch ether is synthesized through the reaction of native starch with alcohol in an acidic medium using a process called etherification (Figure 1) [24–26].

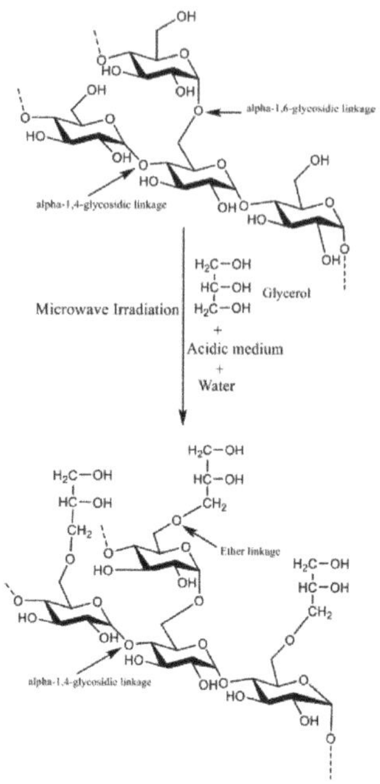

Figure 1. Scheme for the synthesis of partially etherified starch [26].

Starch-based films are thin, transparent films made from starch and other biopolymers, and are used in packaging materials, paper coatings, and other applications. Recently, starch-based thin films have gained increasing attention over traditional synthetic films due to their unique properties such as low cost, high transparency, biodegradability [20,25], renewability, and excellent mechanical properties. They have received great attention due to their sustainable and environmentally friendly characteristics [26] and have been proven as an alternative to petroleum-based and synthetic coatings, which have become a significant source of pollution and environmental degradation. Such benefits and unique properties of starch-based thin films make them suitable for various applications in different industries [27].

In the current work, we have explored a new and sustainable alternative to preserve and decipher charred documents. In this work, starch, a natural polymer, has been exploited to form a coating material. The process involves the creation of a hydrophobic thin film over the charred document and results in a suitable deciphering agent for the documents without even using any spectroscopic or analytical techniques. In order to adopt green synthetic pathways and a quick mode of synthesis, microwave-assisted synthesis procedures were adopted. The characterization of the synthesized starch analog (coating material) was achieved using attenuated total reflectance–Fourier transform infrared spectroscopy (ATR-FTIR) [28], and the effect of the physical properties of the synthesized analog on the mechanical strength of coated charred documents was tested with respect to non-coated charred documents. This work also validated the charred document decipherment using an easily available handwritten character recognition approach (HCR) [29] via Google Lens (G-lens). The present study holds great importance in the field of forensic document examination as well as historical document conservation. In forensic science, the application of starch-based coating for the preservation and decipherment of charred documents and for the retrieval of valuable content enables the use of these preserved and stabilized fragile charred documents as a piece of evidence in a court of law, related to any offense. On the other hand, document conservators can use this technology to preserve historical documents.

2. Materials and Methods

In this study, soluble laboratory starch (extra pure), manufactured by Sisco Research Laboratories Pvt. Ltd. (SRL, Delhi Depot, India), glycerol, acetic acid, and distilled water were used, and the research was conducted in the Department of Forensic Science, Galgotias University.

2.1. Sample Preparation

Charred Samples

Samples were made using a Linc Pentonic brand blue ballpoint pen on 3 different types of papers, viz., 75 g/m^2 A4 size white JK copier paper, 80 g/m^2 A4 size white JK copier paper, and bond paper used for affidavits (e-stamp). Each type of paper was cut into four equal parts (15 × 10.5 cm) and 10 samples of each paper type were prepared, generating a total of 30 samples. A predetermined paragraph was written by the same person on each sample, since different people apply different pressure, which may impact the results.

The prepared samples were charred in a muffle furnace (Thermotech, Faridabad 121001, Haryana) [10] at a temperature ranging from 280 to 310 °C and then removed and placed in a box for safekeeping. The documents charred at temperatures below 300 °C had written contents visible on them. Different grades of paper reach maximum charring with invisible texts at different temperatures (Table 1). Therefore, the samples were charred to the point when they became dark brown to black-grey with the writing completely invisible and were too fragile to handle. In the case of 75 g/m^2 copier paper, it reached maximum charring at 300 °C, 80 g/m^2 copier paper charred at 302 °C, and the bond paper at 305 °C, as shown in Figure 2 In each case, below this temperature, the sample was not appropriately charred, and the writing was visible, and above this temperature, the samples started igniting and gradually turning into ashes. Each sample had its picture taken using a mobile camera (OnePlus Nord CE 5G) for a before and after comparison.

Table 1. Effect of temperature on different types of paper.

Paper Type	Temperature (°C)	Color and Effect on Paper	Effect on Written Text
75 g/m^2 A4 size white JK copier	280	Light brown to white	Visible
	290	Brown in major areas while light brown in some places	Faintly visible in some places
	300	Dark brown to black with slightly curly edges	Completely invisible
	Above 300	Black to grey with gradual ignition from edges, turned into ashes	Completely invisible
80 g/m^2 A4 size white JK copier	290	Light brown to white in a half–half area	Visible
	300	Brown in the major area	Faintly visible in some places
	301	Completely brown in full area	Invisible
	302	Dark brown to black with curly edges	Completely invisible
	Above 302	Black to grey with ignition from edges, turning into grey ashes	Completely invisible
A4 size bond paper Affidavit (e-stamp)	300	Light brown to white in half the area	Visible
	302	Brown in the major area	Faintly visible
	304	Dark brown in full area	Very faintly visible
	305	Dark brown to black	Completely invisible
	Above 305	Black to grey, ignited, turned into ashes	Completely invisible

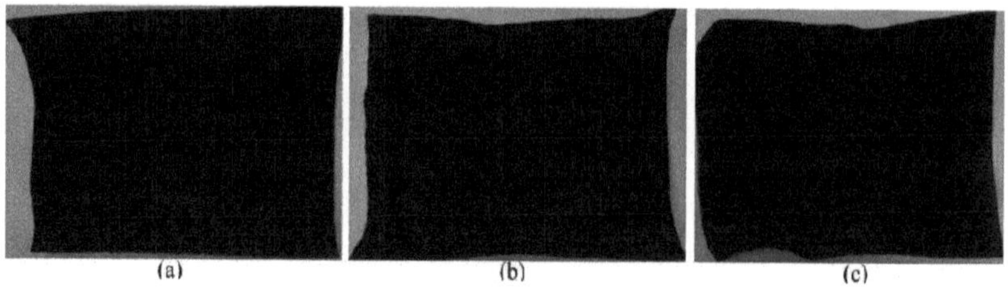

Figure 2. Charred samples: (**a**) 75 g/m^2 (300 °C), (**b**) 80 g/m^2 (302 °C), and (**c**) bond paper (305 °C) made with Linc Pentonic blue ballpoint pen showing invisible and unreadable texts.

2.2. Method of Preservative Preparation

Microwave Synthesis of Starch Analog

Varied-concentration starch solutions (2%, 4%, 6%, and 8%) [30] were made by mixing 2 g, 4 g, 6 g, and 8 g of starch in 100 mL distilled water, respectively, and adding 1 mL glycerol and 1 mL acetic acid to each [31] to optimize the appropriate concentration, as shown in Table 2. The relation between starch and glycerol is approximately in the ratio 1:3. The advantage of using a dilute solution of acetic acid is the low environmental impact compared with other acid solutions [31].

Table 2. Optimization of starch analog with the observed properties.

Concentration (*w*/*v*)	Result
2%	Transparent, low viscosity, not providing appreciable coating, strength, and decipherment
4%	Transparent, viscosity not up to mark, not providing appreciable strength and decipherment
6%	Transparent, appreciable viscosity was suitable for the application, providing appreciable strength and decipherment
8%	Opaque and cloudy, high viscosity, white flaky appearance, not suitable for application and decipherment, cracks observed after drying

Each starch solution was then subjected to microwave irradiation at a temperature of 80 °C and power of 462 watts for about 10 min with constant stirring in the middle until a clear, transparent, sticky analog of starch was acquired.

2.3. Spectral Characterization

The reactants, pure starch powder, glycerol, acetic acid, and the synthesized starch analog, were subjected to ATR-FTIR (Bruker) for spectral characterization to test the presence of different band stretching and the possible formation of the new analog peak after the reaction of starch and glycerol molecules in the acidic medium. The Bruker ATR-FTIR was equipped with a diamond ATR crystal. Prior to the sample analysis, a background spectrum was taken to account for any instrumental or environmental noise by setting the desired measurement conditions (wavenumber 4000–400 cm^{-1}) similar to the sample analysis. Thereafter, the samples were analyzed to acquire their spectral measurements. The spectrum was recorded on the basis of the percentage transmittance (%T) against the wavenumber (cm^{-1}). The measurement was repeated thrice to ensure the reproducibility of the result.

2.4. Application of Starch Analog

Each prepared starch analog was carefully applied over each sample paper type by pouring about 0.5 mL of starch analog over the charred document and spreading it with the help of a small handheld wiper. The wiper used in the application of the starch analog was sufficiently flat, having a smooth silicon base to apply a smooth, evenly coated layer of preservative to the paper surface. The remaining charred samples of each paper type were kept aside for the purpose of comparing their physical properties and the visibility of texts before and after application in coated and non-coated samples.

2.5. Handwriting Character Recognition (HCR)

Soon after the application of the synthesized analog, images of the coated samples were captured using a OnePlus Nord CE 5G mobile camera. Thereafter, they were subjected to handwriting character recognition using Google Lens on the same mobile phone. The captured images were then searched and converted into readable and copiable texts.

2.6. Video Spectral Comparator (VSC)

The samples were visualized under an advanced optical instrument: a video spectral comparator, which incorporates various light sources, widely employed for document examination in forensic science laboratories. The samples were placed in the chamber of the VSC and then were visualized under VIS light, flood light, and white spot light at varied long passes to achieve the maximum visibility of texts.

2.7. Mechanical Properties

2.7.1. Paper Folding Test

The mechanical properties of the coated and non-coated charred documents were preliminarily tested by estimating the increase in the strength of the coated charred documents compared with that of the fragile non-coated charred documents. After the coated charred samples had dried completely, they were subjected to a paper folding test to check the strength. Firstly, the non-coated charred samples were folded at the edges along the sides, and then the same technique was followed with samples coated with the synthesized analog.

2.7.2. Bursting Strength Test

The confirmatory test for the increase in strength of the coated and non-coated charred documents was performed using a Pacorr digital bursting strength tester. This instrument gives quantitative data on the strength of a sample. The bursting strength was calculated based on the force applied by the plunger to the sample under examination, causing it to rupture. Similarly, each non-coated and charred document coated with 6% starch analog was tested to obtain a before-and-after comparison and measure the increased stability.

3. Results and Discussion

In an attempt to preserve and decipher charred documents, it is crucial to note the fact that different documents vary in their physical and chemical composition based on the variety of paper and ink used to make such documents, and also upon the conditions in which they became charred [32]. Therefore, the results of the present research were based on the thickness, i.e., gram per square meter (g/m^2), of the paper, the type of ink used to write the statement, and the coating material used to preserve and decipher the charred document.

As shown in Figure 2, the samples made using different paper types reached maximum charring at different temperatures. The 75 g/m^2 paper reached maximum charring at 300 °C, 80 g/m^2 at 302 °C, and bond paper at 305 °C. Once charred, the documents had invisible text and were too fragile and brittle enough to handle. Thus, they were stabilized and preserved using the synthesized starch analog.

The spectral characterization using ATR-FTIR of reactants (starch powder, glycerol, and acetic acid) and the synthesized analog are shown in Figure 3. Figure 3a shows the ATR-FTIR spectra of pure starch powder, which corresponds to α-1,4-glycosidic linkage at 1143.78 cm^{-1} [33] and α-1,6-glycosidic linkage at 990.79 cm^{-1} and 851.51 cm^{-1} [34], which represent the skeletal model of the amylopectin starch ring and the major -C-OH stretching of the primary and secondary alcoholic group at 3748.08 cm^{-1} and 3271.32 cm^{-1}, respectively (Table 3).

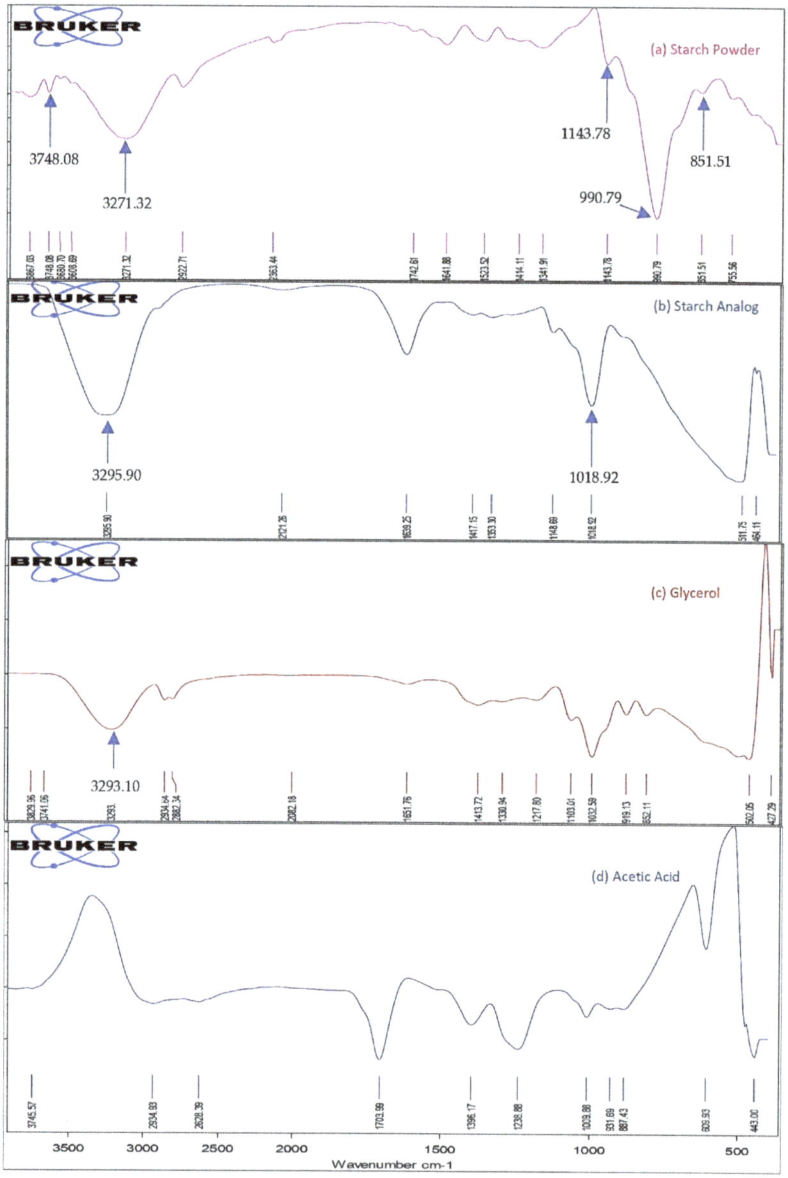

Figure 3. ATR-FTIR spectra of (**a**) starch powder, (**b**) synthesized starch analog, (**c**) glycerol, and (**d**) acetic acid.

The spectral peak at 1018.92 cm^{-1} in Figure 3b indicates the formation of ether linkages (C-O-C) in the starch analog other than glycosidic bonds with the involvement of the C3 hydroxyl group of glycerol and one of the C6 carbons of the monosaccharide moieties of starch. After the reaction between starch and glycerol in the acidic medium, the hydroxyl (-OH) stretching in the starch analog shifted from 3271.32 cm^{-1} to 3295.90 cm^{-1}, and a band of glycosidic bonds shifted from 990.79 cm^{-1} to 1018.92 cm^{-1}, which corresponds to C-O-C stretching and confirms the formation of etherified starch [35–37]. Hence, the result from the ATR-FTIR spectra of the synthesized starch analog shows that partial etherification occurred during a condensation reaction between starch and glycerol. The amount taken for starch and glycerol is in the ratio of 1:3, with glycerol being the limiting agent that carries out partial etherification of starch [38].

Table 3. Major spectral stretching with their wavenumber.

Bond Stretching ↓	Wavenumber (cm^{-1})
Starch Powder	
α-1,4-glycosidic	1143.78
α-1,6-glycosidic	990.79 & 851.51
1°-OH	3748.08
2°-OH	3271.32
Starch Analog	
-OH	3295.90
C-O-C Stretch	1081.92
Glycerol	
-OH	3293.10

In arson cases, documents charred at high temperatures are dehydrated and become brittle and weak. Moreover, the texts on them also vanish due to the evaporation of the volatile solvents of the ink composition, which makes the writings on documents invisible. However, the residue of inks components like fats and oils may still be present on the charred documents but invisible. The disappearance of written texts and the blackening of documents due to excessive heat and smoke become a serious problem for the document under decipherment [39]. On application of the optimized starch analog, it was found that in normal daylight, soon after the application, the texts became visible to the naked eye in most of the places of the charred document, which were previously, i.e., before preservative application, invisible to the naked eye as well as in oblique lighting. This may be due to the ballpoint pen ink composition and due to the removal of carbon and smoke particles embedded in the grooves made by the pen pressure on the charred samples.

Figure 4 shows the coated charred documents with 2% starch analog. After applying starch analog, it did not give appreciable results as the hydrophobic coating did not form on the samples and was not dry even after exposure to the air for hours. Once it dried, a thin film formed, which could not provide appreciable strength to the charred samples, as tested by the paper folding test. The decipherment of text was also not sufficient as none of the text could be read by the naked eye in the case of the bond paper coated with 2% analog, while only a few letters of the statement were read on the 75 g/m^2 and 80 g/m^2 paper. The same results were confirmed using HCR via Google Lens (G-lens) (Figure 5). This may be due to the lower viscosity of the analog, which could not provide enough coating and hence the decipherment of the invisible texts. Similarly, the application of 4% starch analog (Figure 6) also did not provide promising results, as only a few texts could be read in the case of the 80 g/m^2 and bond paper coated with 4% analog, while the majority of texts were deciphered on the 75 g/m^2 paper, which was confirmed using the HCR technique through G-lens (Figure 7).

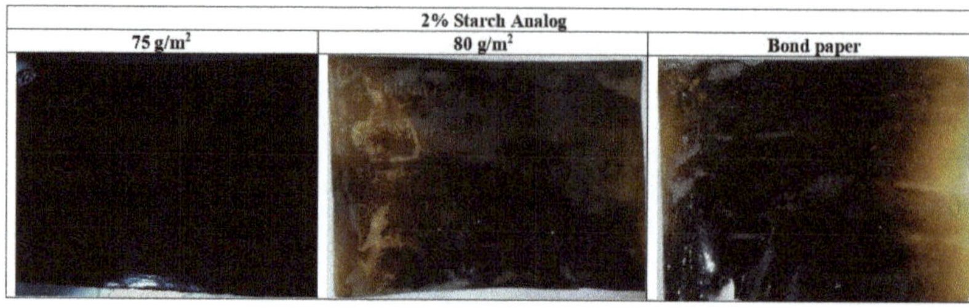

Figure 4. Coated charred samples with 2% starch analog on 75 g/m^2, 80 g/m^2, and bond paper (affidavit) written with Linc Pentonic blue ballpoint pen.

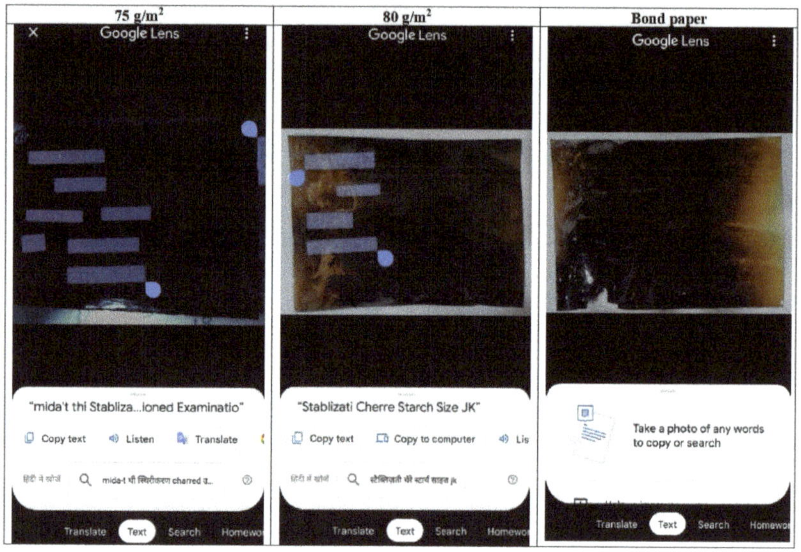

Figure 5. HCR-read texts of coated charred samples with 2% starch analog on 75 g/m^2, 80 g/m^2, and bond paper (affidavit).

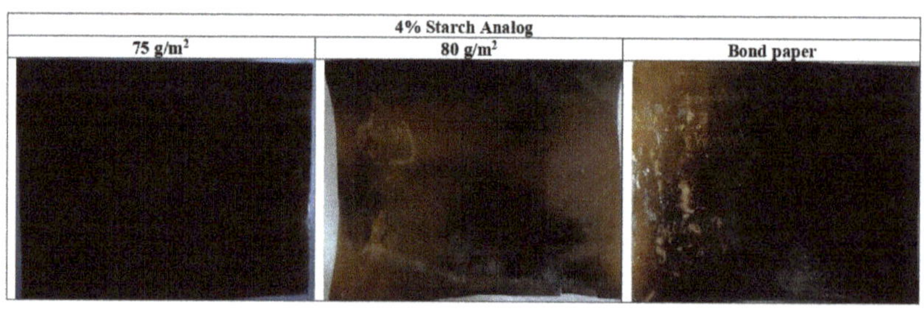

Figure 6. Coated and preserved charred samples with 4% starch analog on 75 g/m^2, 80 g/m^2, and bond paper (affidavit) written with Linc Pentonic blue ballpoint pen.

Moreover, 8% starch analog was found to be viscous enough to form an even coating on the charred document that resulted in the formation of white precipitated flakes of starch, which hindered smooth application. None of the text could be deciphered on the 75 g/m^2 and bond paper, while only some could be faintly seen on the 80 g/m^2 paper through the naked eye (Figure 8), as well as using G-lens (Figure 9). After it had dried, cracks of coating developed on the surface, and it curled.

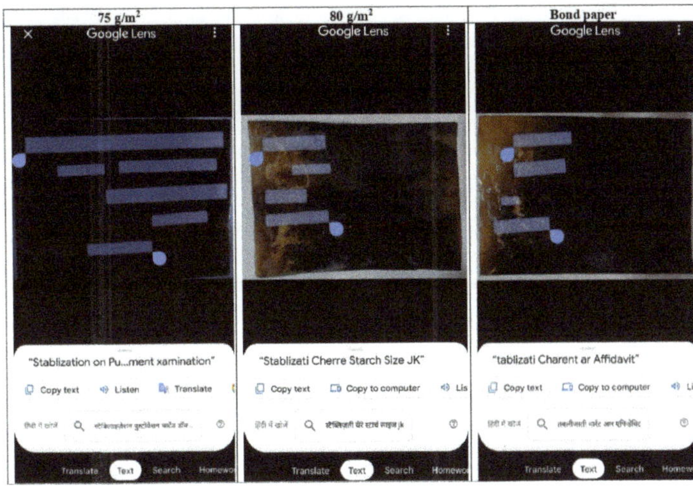

Figure 7. HCR-read texts of coated charred samples with 4% starch analog on 75 g/m^2, 80 g/m^2, and bond paper (affidavit).

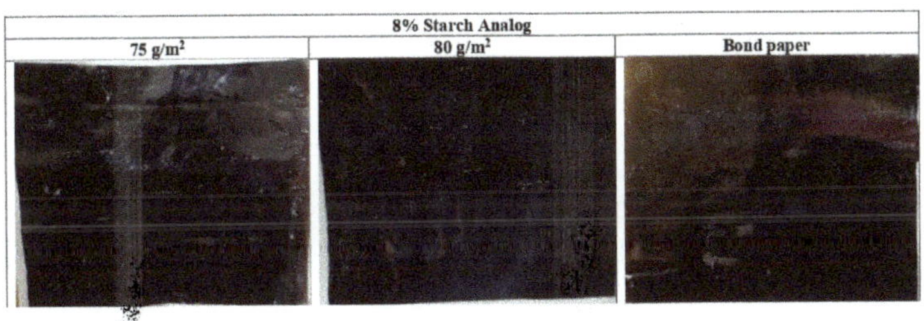

Figure 8. Coated and preserved charred samples with 8% starch analog on 75 g/m^2, 80 g/m^2, and bond paper (affidavit) written with Linc Pentonic blue ballpoint pen.

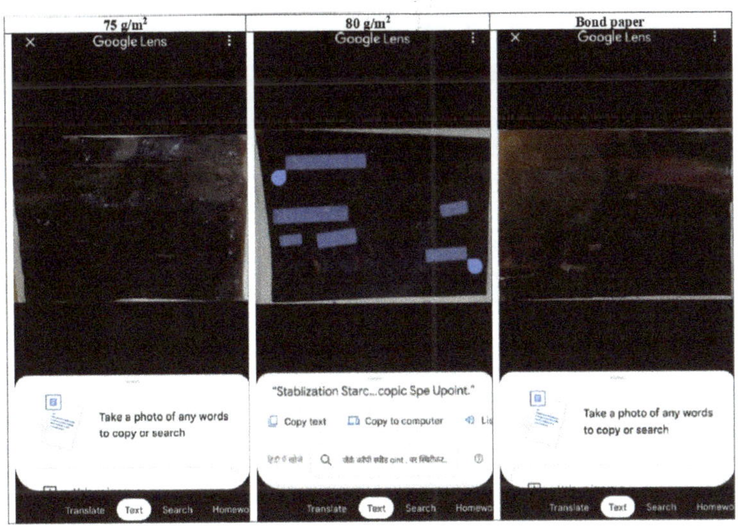

Figure 9. HCR-read texts of coated charred samples with 8% starch analog on 75 g/m^2, 80 g/m^2, and bond paper (affidavit).

In contrast, Figure 10 shows the application of 6% starch analog. The hydrophobic film formed like a polish that smoothed the surface and reduced the fragility and brittleness of the document through the process of rehydration. Soon after application, the texts were deciphered and visible to the naked eye. This may be due to the difference in polarity of the starch analog, consisting of glycerol, and ink composition, consisting of oils and fats. The residual traces of oils and fats on the burnt paper act as a repellent to the synthesized analog. Hence, the starch analog is absorbed by the charred paper except on the writing, leading to the differentiation due to color contrast between the writing and the background [10,40]. The HCR technique using G-lens also recognized the maximum deciphered text of the statement on all three types of paper, i.e., 75 g/m^2, 80 g/m^2, and bond paper (Figure 11). The coated charred document also dried in about 10–15 min at room temperature, which gave an extra advantage of this synthesized starch analog.

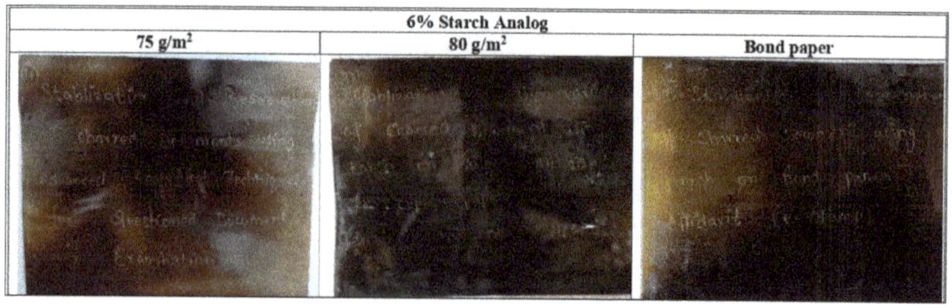

Figure 10. Coated and preserved charred samples with 6% starch analog on 75 g/m^2, 80 g/m^2, and bond paper (affidavit) written with Linc Pentonic blue ballpoint pen.

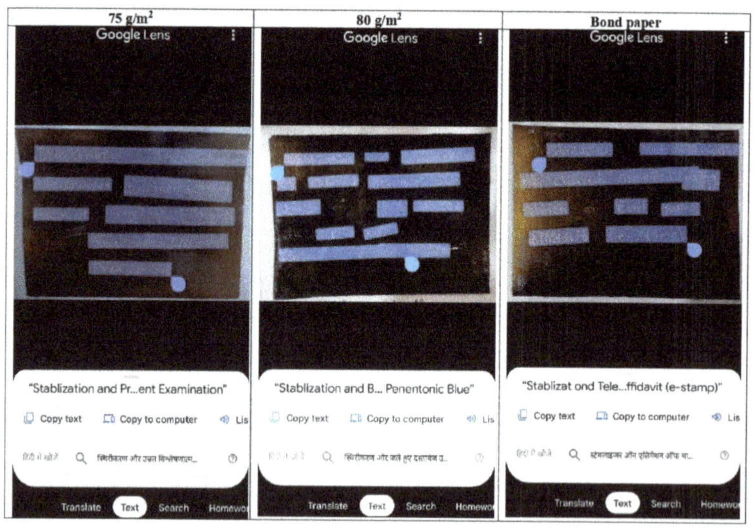

Figure 11. HCR-read texts of coated charred samples with 6% starch analog on 75 g/m^2, 80 g/m^2, and bond paper (affidavit).

The further analysis of the coated charred documents under an advanced optical VSC instrument with different light sources and at varied long passes gave quite good results of decipherment even a month after coating the charred document using starch analog. In the case of the 75 g/m^2 and 80 g/m^2 charred documents written with blue ballpoint pen ink, the texts were enhanced and deciphered under flood light at 715 long pass, whereas on the bond paper, the texts were more prominently legible under the white spot light at 695 long pass, as shown in Figure 12. This is due to the reflection of the different wavelengths of light from the document surface under visualization, which makes the invisible and low-visibility text prominently visible.

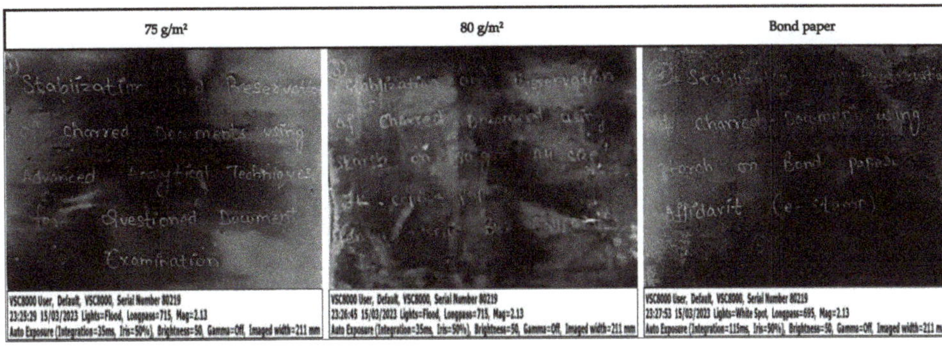

Figure 12. VSC-read texts of coated charred samples with 6% starch analog on 75 g/m^2, 80 g/m^2, and bond paper (affidavit).

Moreover, the mechanical properties of the coated charred documents showed that the coatings formed on the charred documents by applying the starch analog increased the strength of the fragile charred documents, which was tested preliminary by the paper folding test. Further, an experimental, qualitative analysis using a digital bursting strength tester calculated the bursting strength of the coated and non-coated charred documents, which provided promising results. The bursting strength of the 75 g/m^2 non-coated charred documents before stabilization was found to be 0.12 kg/cm^2, which increased to 0.21 kg/m^2 after stabilization with starch analog, whereas the bursting strength of the non-coated 80 g/m^2 and bond paper was 0.15 kg/cm^2 and 0.25 kg/cm^2, which increased to 0.25 kg/cm^2 and 0.35 kg/cm^2 after being stabilized with starch analog, as shown in Figure 13 [41,42]. This may be due to the penetration and absorbance of the starch analog on the charred layer of paper, which adhered to the surface, thus strengthening the fragile and brittle charred documents, which can be preserved and placed in safekeeping for a few more days.

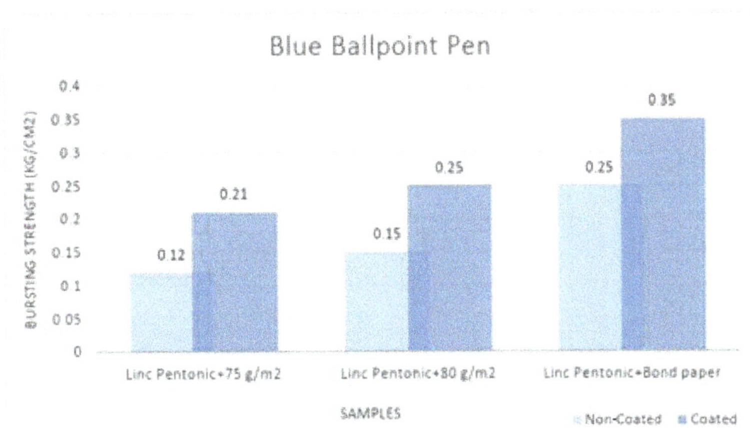

Figure 13. Graphical representation of the increase in bursting strength between non-coated and coated charred documents made using Linc blue ballpoint pens on 75 g/m^2, 80 g/m^2, and bond paper.

4. Conclusions

The present research concludes that the preservation and decipherment of texts of charred documents by any means depends on the type of paper, writing ink, condition, and level of charring. Soon after the application of 6% starch analog, the texts were visible on all three types of paper samples charred at varied temperatures. This gave the best result, since the texts could also be clearly recognized through handwriting character recognition (HCR) via Google Lens to check the readability level by converting the images to readable and copiable text. Moreover, it was found that the drying and setting (time) of the synthesized analog after application on charred samples took 10–15 min only. In addition, a protective layer of polysaccharide provided an increase in mechanical strength to the charred samples, which was observed and tested by general observation of thickness and the folding test method comparing coated and non-coated charred samples. Moreover, the sample dried, turned into a hydrophobic film, and was not affected by water, thus making the document moisture-protected. Hence, it can be concluded that for the documents made with a ballpoint pen preserved with 6% starch analog if charred up to around 300 °C (75 g/m^2), 302 °C (80 g/m^2), and 305 °C (bond paper affidavit), the writing can be deciphered and read easily, and the method is easy, non-toxic, and cost-effective. The present study suggests exploring the results based on more combinations of pens like gel pens, fountain pens, markers, etc., on different grades and colors of paper that are usually used in crucial documents. By delving into more combinations of writing instruments such as gel pens, fountain pens, etc., studies can expand their scope and investigate their impact on the preservation and decipherment of charred documents. This broader approach will provide a comprehensive understanding of how various writing tools interact with different grades and colors of paper during the charring process, particularly in the context of crucial documents. The outcomes of this extended research could potentially lead to the development of enhanced techniques and strategies for recovering and interpreting vital information from charred documents, thereby advancing the field of document analysis and historical research. Furthermore, these findings may have practical applications in forensic investigations and archival preservation, ultimately contributing to the preservation and understanding of invaluable historical records.

Author Contributions: Conceptualization, S.K. (Sonali Kesarwani) and D.B.T.; methodology S.K. (Sonali Kesarwani); software, S.K. (Sonali Kesarwani). and D.B.T.; validation, S.K. (Sonali Kesarwani) and D.B.T.; formal analysis, S.K. (Sonali Kesarwani).; investigation, S.K. (Sonali Kesarwani). and D.B.T.; resources, S.K. (Suneet Kumar).; data curation, S.K. (Sonali Kesarwani); writing—original draft preparation, S.K. (Sonali Kesarwani) and D.B.T.; writing—review and editing, S.K. (Sonali Kesarwani) and D.B.T.; visualization, S.K. (Sonali Kesarwani); supervision, D.B.T. and S.K. (Suneet Kumar); project administration, D.B.T. All authors have read and agreed to the published version of the manuscript.

Funding: This research received no external funding.

Institutional Review Board Statement: Not applicable.

Informed Consent Statement: Not applicable.

Data Availability Statement: No additional data are available.

Acknowledgments: The authors acknowledge the contribution of Anurag Singh, HBTU, Kanpur, and Vivek Kumar Bajpai, Department of Energy and Material Science Engineering, Dongguk University, Republic of Korea, for his additional guidance.

Conflicts of Interest: The authors declare no conflict of interest.

References

1. Cronstedt, M. Prevention, preparedness, response, recovery-an outdated concept? *Aust. J. Emerg. Manag.* **2002**, *17*, 10–13.
2. Blagden, C. Some observations on ancient inks, with the proposal of a new method of recovering the legibility of decayed writings. *Philos. Trans. R. Soc. Lond.* **1787**, *77*, 451–457. [CrossRef]

3. Davis, R. *Action of Charred Paper on the Photographic Plate and a Method of Deciphering Charred Records*; US Government Printing Office: Washington, DC, USA, 1922; p. 18.
4. Mitchell, C.A. The examination of charred documents. *Analyst* **1925**, *50*, 174–180. [CrossRef]
5. Mitchell, C.A. The use of infra-red rays in the examination of inks and pigments. *Analyst* **1935**, *60*, 454–461. [CrossRef]
6. Radley, J.A.; Grant, J. Ultra-Violet Light in Technology. *Nature* **1940**, *146*, 216. [CrossRef]
7. Donald, D. Charred Documents, their Handling, and Decipherment: A Summary of available Methods for Treating Burnt Papers. *J. Crim. L Criminol. Police Sci.* **1952**, *43*, 812.
8. Chayal, V.M.; Paite, N.L.; Barik, A.K.; Kumar, S.; Bhattacharya, S.; Amin, H. Stabilization and Examination of Charred Documents: A Systemic Review. *J. Forensic Res.* **2023**, *1*, 14.
9. Simon, H.; David, R. Stabilisation of charred documents using alkyl-2-cyanoacrylate ester (superglue fuming). *Aust. J. Forensic Sci.* **2013**, *45*, 103–105.
10. Tidke, R.C.; Kushwaha, K.P.S. Decipherment of Ball Point Pen Writings on the Charred/Burnt Paper. *Int. J. Forensic Sci.* **2019**, *2*, 1.
11. Wilson, H.R. *Suspect Documents, Their Scientific Examination*; Nelson-Hall: Raleigh, NC, USA, 1981. Available online: https://cir.nii.ac.jp/crid/1130282272630591232 (accessed on 22 May 2023).
12. Carney, B.R. A charred Document Case Made Simple: Methods for the examination and protection of charred document evidence. *Int. J. Forensic Doc. Exam.* **1996**, *1*, 347–348.
13. Gadhave, R.V.I. Starch Grafted Water Resistant Polyvinyl Acetate-Based Wood Adhesive: A Review. *Open J. Org. Polym. Mater.* **2022**, *12*, 17–30. [CrossRef]
14. de Almeida Assis, A.C.; da Fonseca, J.F.; de Fátima Barbosa, M.; Farinha, C. Forensic Analysis of Unknown Materials: A different vision of questioned documents. *Eur. L Enforc. Res. Bull.* **2018**, *17*, 67.
15. Altuna, L.; Herrera, M.L.; Foresti, M.L. Synthesis and characterization of octenyl succinic anhydride-modified starches for food applications. A review of recent literature. *Food Hydrocoll.* **2018**, *80*, 97–110. [CrossRef]
16. Jansson, A.; Järnström, L. Barrier and mechanical properties of modified starches. *Cellulose* **2005**, *12*, 423–433. [CrossRef]
17. Xu, J.; Andrews, T.D.; Shi, Y. Recent Advances in the Preparation and Characterization of Intermediately to Highly Esterified and Etherified Starches: A Review. *Starch-Stärke* **2020**, *72*, 3–4. [CrossRef]
18. Li, Q.; Wang, S.; Jin, X.; Huang, C.; Xiang, Z. The Application of Polysaccharides and Their Derivatives in Pigment, Barrier, and Functional Paper Coatings. *Polymers* **2020**, *12*, 1837. [CrossRef]
19. Zuo, Y.; Gu, J.; Yang, L.; Qiao, Z.; Tan, H.; Zhang, Y. Preparation and characterization of dry method esterified starch/polylactic acid composite materials. *Int. J. Biol. Macromol.* **2014**, *64*, 174–180. [CrossRef]
20. Lu, D.R.; Xiao, C.M.; Xu, S.J. Starch-based completely biodegradable polymer materials. *Express Polym. Lett.* **2009**, *3*, 366–375. [CrossRef]
21. Hassan, R.R.A.; Mahmoud, S.M.A.; Karam, Y.A.; Salah, S.M.; Ebrahim, S.Y.; Abdelwahab, M.A.; Ahmed, A.-H.M.H.; Ali, H.M.; Böhm, M.; Salem, M.Z.M. Application of frankincense and rice starch as eco-friendly substances for the resizing of paper as a conservation practice. *BioResources* **2021**, *16*. [CrossRef]
22. Maurer, H.W. Starch in the Paper Industry. In *Starch*, 3rd ed.; Miller, J.B., Whistler, R., Eds.; Academic Press: San Diego, CA, USA, 2009; pp. 657–713.
23. Maurer, H.W. Use of Starch for Surface Sizing of Paper. In *Starch*, 3rd ed.; Miller, J.B., Whistler, R., Eds.; Academic Press: San Diego, CA, USA, 2009; pp. 688–695.
24. Wang, X.; Huang, L.; Zhang, C.; Deng, Y.; Xie, P.; Liu, L.; Cheng, J. Research advances in chemical modifications of starch for hydrophobicity and its applications: A review. *Carbohydr. Polym.* **2020**, *240*, 116292. [CrossRef] [PubMed]
25. Otache, M.A.; Duru, R.U.; Achugasim, O.; Abayeh, O.J. Advances in the Modification of Starch via Esterification for Enhanced Properties. *J. Polym. Environ.* **2021**, *29*, 1365–1379. [CrossRef]
26. Seligra, P.G.; Jaramillo, C.M.; Famá, L.; Goyanes, S. Biodegradable and non-retrogradable eco-films based on starch–glycerol with citric acid as crosslinking agent. *Carbohydr. Polym.* **2016**, *138*, 66–74. [CrossRef] [PubMed]
27. Sharma, S.; Garg, D.; Chophi, R.; Singh, R. On the spectroscopic investigation of stamp inks using ATR-FTIR and chemometrics: Application in forensic document examination. *Forensic Chem.* **2021**, *26*, 100377. [CrossRef]
28. Li, Z.; Liu, X. An examination of handwritten signatures forged using photosensitive signature stamp. *Forensic Sci. Res.* **2021**, *6*, 168–182. [CrossRef]
29. Memon, J.; Sami, M.; Khan, R.A.; Uddin, M. Handwritten Optical Character Recognition (OCR): A Comprehensive Systematic Literature Review (SLR). *IEEE Access* **2020**, *8*, 142642–142668. [CrossRef]
30. Domene-López, D.; García-Quesada, J.C.; Martin-Gullon, I.; Montalbán, M.G. Influence of Starch Composition and Molecular Weight on Physicochemical Properties of Biodegradable Films. *Polymers* **2019**, *11*, 1084. [CrossRef] [PubMed]
31. Montecinos Espinoza, S.D.L.A.; Ascazuri, M.; Achaga, J.; Viduzzi, G.; Tognana, S.A. Biopolymers to mitigate the environmental impact. *Inglomayor Mag.* **2019**, *17*, 775.
32. Tyrrell, J.F. The Decipherment of Charred Documents. *J. Crim. Law Criminol.* **1939**, *30*, 236. [CrossRef]
33. Nikonenko, N.A.; Buslov, D.K.; Sushko, N.I.; Zhbankov, R.G. Investigation of stretching vibrations of glycosidic linkages in disaccharides and polysaccharides with the use of IR spectra deconvolution. *Biopolym. Orig. Res. Biomol.* **2000**, *57*, 257–262.
34. Synytsya, A.; Miroslav, N. Structural analysis of glucans. *Ann. Transl. Med.* **2014**, 22. [CrossRef]

35. Paluch, M.; Ostrowska, J.; Tyński, P.; Sadurski, W.; Konkol, M. Structural and Thermal Properties of Starch Plasticized with Glycerol/Urea Mixture. *J. Polym. Environ.* **2021**, *30*, 728–740. [CrossRef]
36. Ojogbo, E.; Ogunsona, E.O.; Mekonnen, T.H. Chemical and physical modifications of starch for renewable polymeric materials. *Mater. Today Sustain.* **2020**, *7*, 100028. [CrossRef]
37. Zhang, Y.; Han, J.H. Plasticization of Pea Starch Films with Monosaccharides and Polyols. *J. Food Sci.* **2006**, *71*, E253–E261. [CrossRef]
38. Masina, N.; Choonara, Y.E.; Kumar, P.; du Toit, L.C.; Govender, M.; Indermun, S.; Pillay, V. A review of the chemical modification techniques of starch. *Carbohydr. Polym.* **2017**, *157*, 1226–1236. [CrossRef] [PubMed]
39. Kesarwani, S.; Tripathy, D. Advancements in Potential Preservation and Decipherment Techniques of Charred Documents. *Indian J. Forensic Med. Pathol.* **2021**, *2*, 359–366.
40. Nabilah, M.R.N.; Alwi, M.A.; Su'ait, M.S.; Imperiyka, M.; Hanifah, S.A.; Ahmad, A.; Rahman, M.Y.A. Effect of ionic liquid 1-butyl-3-methylimidazolium bis (trifluoromethane sulfonyl) imide on the properties of poly (glycidyl methacrylate) based solid polymer electrolytes. *Russ. J. Electrochem.* **2016**, *52*, 362–373. [CrossRef]
41. Li, H.; Qi, Y.; Zhao, Y.; Chi, J.; Cheng, S. Starch and its derivatives for paper coatings: A review. *Prog. Org. Coat.* **2019**, *135*, 213–227. [CrossRef]
42. Biricik, Y.; Sonmez, S.; Ozden, O. Effects of surface sizing with starch on physical strength properties of paper. *Asian J. Chem.* **2011**, *23*, 3151.

coatings

Communication

Systematic Investigation into Evolution of Materials and Techniques Used in Lacquer Lian from the Warring States Period to the Yuan Dynasty

Hao Wu [1], Yang Zhao [1], Beisong Fang [1] and Jingren Dong [2,*]

[1] Jingzhou Conservation Institute, Jingzhou 434020, China
[2] College of Materials Science and Engineering, Chongqing University, Chongqing 400044, China
* Correspondence: jingrendong@sina.com

Abstract: In order to investigate the evolution of Chinese lacquering techniques, seven pieces of lacquer Lian from the Warring States Period to the Yuan Dynasty (475 BC–1368 AD) were analyzed by means of cross-section observation, Raman spectroscopy (RS), and thermally assisted hydrolysis and methylation pyrolysis coupled with gas chromatography/mass spectroscopy (Py-GC/MS). The results revealed that the lacquer Lian consisted of a three-layer structure, encompassing a pigment layer on the surface, an undercoat layer in the middle, and a ground layer. The red mineral pigment utilized was cinnabar, while a combination of Chinese lacquer and drying oil served as the primary organic material. Although lacquering techniques had undergone minimal changes from the Warring States Period to the Yuan Dynasty, the species of drying oil had changed, based on the fact that boiled tung oil was found in the ground layer of lacquerware from the Song Dynasty and the Yuan Dynasty. The present research provides direct evidence for the inheritance and development of Chinese lacquer technology.

Keywords: Lacquer Lian; Py-GC/MS; Chinese lacquer; boiled tung oil; cross sections; lacquering techniques

Citation: Wu, H.; Zhao, Y.; Fang, B.; Dong, J. Systematic Investigation into Evolution of Materials and Techniques Used in Lacquer Lian from the Warring States Period to the Yuan Dynasty. *Coatings* **2023**, *13*, 1750. https://doi.org/10.3390/coatings13101750

Academic Editor: Maurizio Licchelli

Received: 8 September 2023
Revised: 7 October 2023
Accepted: 8 October 2023
Published: 10 October 2023

1. Introduction

Lacquer, a natural coating material for wood, porcelain, and metal, is tapped from lacquer trees growing in different regions of East and Southeast Asia: Rhus vernicifera (China, Japan, and Korea), Rhus succedanea (Vietnam and Taiwan), and Melanorrhoea usitata (Laos, Burma, Thailand, and Cambodia) [1,2]. The main component of lacquer is a mixture of catechol and phenol derivatives (60%–65%), proteins (glycoproteins (2%) and a laccase enzyme (1%)), polysaccharides (7%), and water (30%) [3]. Lacquer, when catalyzed by laccase, can undergo polymerization in the natural surroundings, resulting in the formation of a film characterized by an intricate three-dimensional network [4,5]. Exhibiting exceptional stability and possessing remarkable abilities to resist water, corrosion, and microbial growth, lacquer emerges as an exemplary environmentally friendly substance.

Lacquered artifacts have been revered and esteemed due to their unparalleled fortitude, enduring resilience, imperviousness, and exquisite beauty across a prolonged span of time. The utilization of lacquer throughout history can be traced back to the Neolithic era, and in the context of China's heritage, it can be classified into six distinct phases. The initial phase is the period of incubation, predominantly during the Neolithic epoch. The subsequent phase is the period of germination, transpiring during the Bronze Age, encompassing the Xia, Shang, and Zhou dynasties. This is followed by the phase of growth, occurring in the Iron Age, specifically during the Chunqiu Zhanguo dynasty. The fourth phase emerges as the heyday, characterized as the era of lacquer, flourishing primarily during the Qin and Han dynasties. The fifth phase, known as the recession stage, arises during the Buddhist era, spanning from the Wei to Tang dynasties. Ultimately, the final stage unveils itself as the pinnacle of prosperity within the porcelain era, comprising the Song, Yuan, Ming, and Qing dynasties [6]. Initial evidence of Chinese lacquerware can be traced back over 8000 years ago to the Kuahuqiao culture, where an ancient wooden bow delicately adorned with raw lacquer was unearthed [7]. A red lacquered wooden bowl was unearthed at Hemudu, a Neolithic site in the Yangtze River Delta, that can be traced back to 6000–7000 years ago [8]. In the present day, lacquerware products continue to hold their esteemed position as the most esteemed and sought-after handicrafts globally.

The red lacquer film commonly includes a subsequent mineral constituent such as iron oxide red (Fe_2O_3), lead red (Pb_3O_4) [9], or cinnabar (HgS). The primary constituents of the yellow lacquer film are orpiment or realgar. The black lacquer film frequently incorporates carbon black [10] or ferroferric oxide. Throughout the annals of lacquerware artistry, the red, yellow, and black colors have assumed significant roles. In this investigation, the main colors observed in the lacquer film specimens were red and black, aligning harmoniously with traditional aesthetics.

As we know, a number of modern analytical methods have been used to characterize lacquerwares, such as optical microscopy (OM) [11], scanning electron microscopy augmented by energy-dispersive X-ray spectrometry (SEM-EDS) [12], Raman spectroscopy [13], Fourier transform infrared spectrometry (FTIR) [14], and pyrolysis–gas chromatography/mass spectrometry (Py-GC/MS) [15]. A large number of scientific studies have been undertaken to characterize Asian lacquer, focusing on molecular markers to identify the three lacquer species [16], the interactions between plant oils and lacquers [17], and the pathways of lacquer degradation [18]. Most research has been aimed at a single item of ancient lacquerware, several in one burial, or several in different burials but from the same period [19–22]. Although China was one of the earliest countries to make and use lacquerware, few studies about its inheritance and evolution of lacquering techniques have been reported. The sole extant treatise on the art of lacquering is *Xiushilu*, authored by Huang Cheng during the Ming Dynasty. Despite providing the groundwork for the exploration of lacquer techniques, the accounts within *Xiushilu* are regrettably concise, limiting our contemporary comprehension of the materials and methodologies employed in the creation of such exquisite lacquerware. Lacquer Lian was an important daily necessity in ancient times for Chinese women to store their toiletries, and is very common in unearthed cultural relics from the Warring States Period to the Tang and Song Dynasties, making it very suitable for research on the inheritance and evolution of Chinese ancient lacquer techniques. In this study, seven lacquer Lian samples from the Warring States Period, Han Dynasty, Song Dynasty, and Yuan Dynasty were analyzed via numerous modern analytical methods to investigate the types of lacquer, pigments, and drying oil added to the lacquer. The inheritance and evolution of Chinese ancient lacquer techniques is discussed by comparing the analysis results of seven lacquerware samples. Moreover, this study can also provide scientific support for the preservation and conservation of unearthed lacquerware.

2. Experimental

2.1. Archaeological Samples

Seven fragment samples of lacquer Lian used in this study were supplied by Jingzhou Museum, Yangzhou Museum, Changzhou Museum, and Jinsha site Museum, respectively. Two of the samples (Samples 1 and 2) dating back to the Warring States Period were unearthed from Jigongshan and Yangjiashan tomb, Hubei Province, respectively. Two samples (Samples 3 and 4) were traced back to Han Dynasty from Tianhui town tomb and Fenghuang mountain tomb, Sichuan Province. Two samples (Samples 5 and 6) dating back to the Song Dynasty were supplied by Yangzhou Museum and Changzhou Museum, respectively. The last one (Sample 7), tracing traced back to the Yuan Dynasty, was provided by Yangzhou Museum. Images of the lacquer fragments are shown Figure 1.

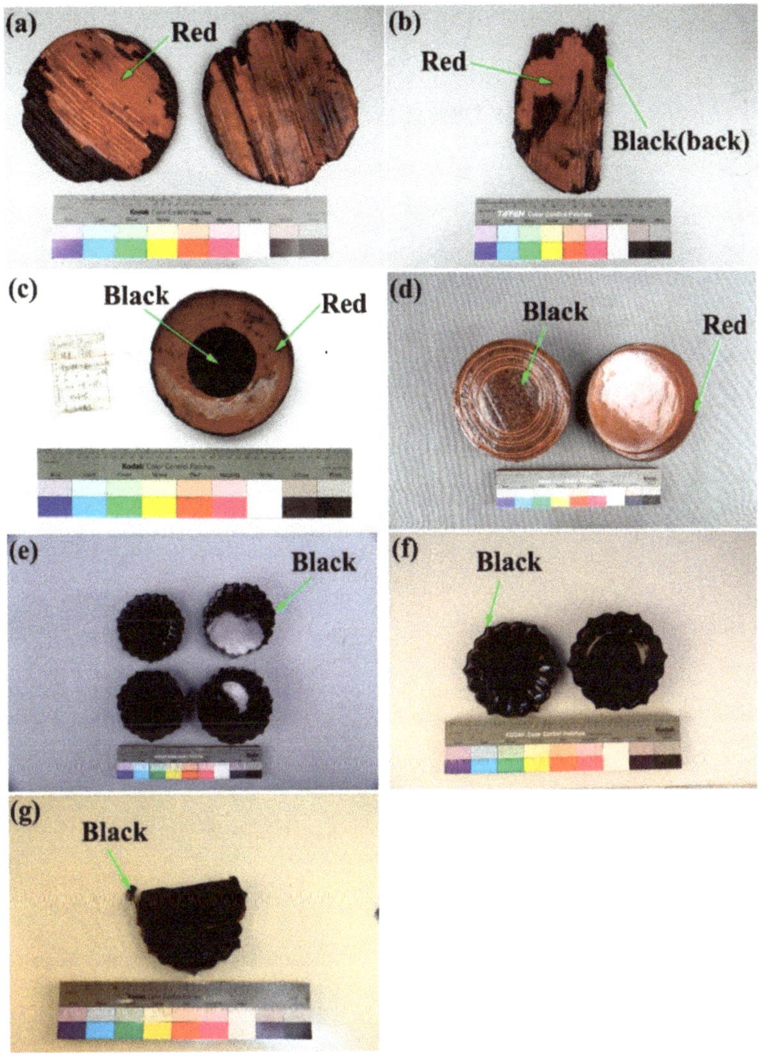

Figure 1. Images of lacquer fragments of (**a**,**b**) Warring States Period, (**c**,**d**) Han Dynasty, (**e**,**f**) Song Dynasty, and (**g**) Yuan Dynasty. The sampling locations are indicated by green arrows.

2.2. Analysis Methods

2.2.1. Cross-Section Observations

Lacquer samples (length × height, 10 mm × 5 mm) were embedded in epoxy resin and the bottom surface was polished with dry sandpaper (up to 12000#) after the epoxy resin was completely cured. The cross-section observations were carried out with an optical microscope (Axio Scope A1, Zeiss, Goettingen, Germany) under blue light (BL). The thickness of each layer was acquired by measuring ten data points using AxioVision software (Version 4.9) and calculating their average value.

2.2.2. Raman Spectroscopy

A micro confocal Raman spectrometer (LabRAM HR Evolution, HORIBA, Loos, France) was used to determine the mineral pigment in the lacquer film layers. The analysis was conducted using an excitation wavelength of 532 nm. The Raman spectra were acquired across the spectral range of 100–800 cm^{-1} with a resolution of 2 cm^{-1}, and subsequently cross-referenced with the RRUFF Raman spectroscopy databases.

2.2.3. Thermally Assisted Hydrolysis and Methylation Pyrolysis Coupled with Gas Chromatography/Mass Spectroscopy (THM-Py-GC/MS)

The pyrolysis–gas chromatography/mass spectroscopy measurements were carried out using a PY-3030D pyrolyzer (Frontier Lab, Fukushima, Japan) attached to a GCMS-QP2020 gas chromatograph mass spectrometer (Shimadzu, Kyoto, Japan). A stainless-steel capillary column (0.25 mm i.d. × 30 m) coated with 0.25 μm of 100% dimethylpolysiloxane was used for separation. Online methylation with additions of less than 1 mg sample and 5 μL 25% aqueous solution of tetramethyl ammonium hydroxide was used to obtain methylated phenolic hydroxyl groups. The sample was pyrolyzed at 500 °C for 0.2 min. The temperature of the pyrolyzer and GC interface was 300 °C. The initial temperature of the gas chromatograph oven was set to 50 °C, held at this temperature for 5 min, and then increased from 50 to 300 °C at 4 °C/min. The temperature of the gas chromatograph oven was then maintained at a constant temperature of 300 °C for 15.5 min. The temperatures of the injector and ion source were set to 300 and 230 °C, respectively. Helium was used as the carrier gas at a flow rate of 1.0 mL/min with a split ratio of 1:20. The electron ionization energy for mass spectroscopy was 70 eV, and the scanning range was from m/z 10 to 600 with full scan mode. Compounds were identified using comparisons between mass spectroscopy and the NIST library.

3. Results and Discussion

3.1. Cross-Sectional Analysis

The morphology of the cross sections of seven samples were examined under optical microscope. As shown in Figure 2, their cross sections showed similar three-layer structures, including a colored paint layer with red pigment or a surface finish layer (Layer 1), an undercoat layer (Layer 2), and a ground layer (Layer 3). Layer 1 boasted a more uniform thickness in comparison to Layer 2. The detailed thickness of the lacquer films of the analyzed archaeological samples is shown in Table 1. The difference in thickness between the red layer and the black layer is distinct, with the black layer being significantly thicker than the red layer (21.6 vs. 19.2, 40.8 vs. 29.9, 39 vs. 27.6). The red colored layer became thicker from the Warring States Period to the Han Dynasty (21.6, 19.2 vs. 29.9, 27.6).

Table 1. The detailed thickness of lacquer films of the analyzed archaeological samples.

Sample	Color	Layer 1 (μm)		Layer 2 (μm)	
		Average	Standard Deviation	Average	Standard Deviation
1	red	21.6	2.89	56.7	8.84
2	black	21.6	1.34	28.1	11.84
	red	19.2	5.28	22.7	10.44
3	black	40.8	6.15	-	-
	red	29.9	0.84	23.0	3.20
4	black	39.0	3.92	11.7	1.74
	red	27.6	2.69	5.1	0.97
5	black	43.8	2.45	78.9	20.67
6	black	31.6	1.94	9.9	4.59
7	black	45.3	5.75	35.1	10.86

The colored paint layer, known to be a mixture of pigment, drying oil, and lacquer liquid, was not only used to decorate the lacquerware items, but also could increase the covering power of the lacquer film, prevent ultraviolet rays from penetrating the lacquer film, and delay the aging of the lacquer film. It is located on the top layer of the lacquerware, as shown in Figure 2a,c,e,g. The undercoat layer under the surface layer, which was usually a mixture of lacquer and drying oil, was used to cover defects on the next layer, such as the second layer in Figure 2, which resulted in uneven thickness. It can also be used as the surface finish layer for the original color of raw lacquer, as shown in Figure 2b,d,f,h–j. A mixture of clay, lacquer, and oil, known as the ground layer, was used to fill the pores of lacquer bodies, as shown in Figure 2. As an important daily necessity in ancient times, the manufacturing process of lacquer Lian involved a three-layer structure, which was a commonly used technique in ancient Chinese lacquerware such as ear cups [19].

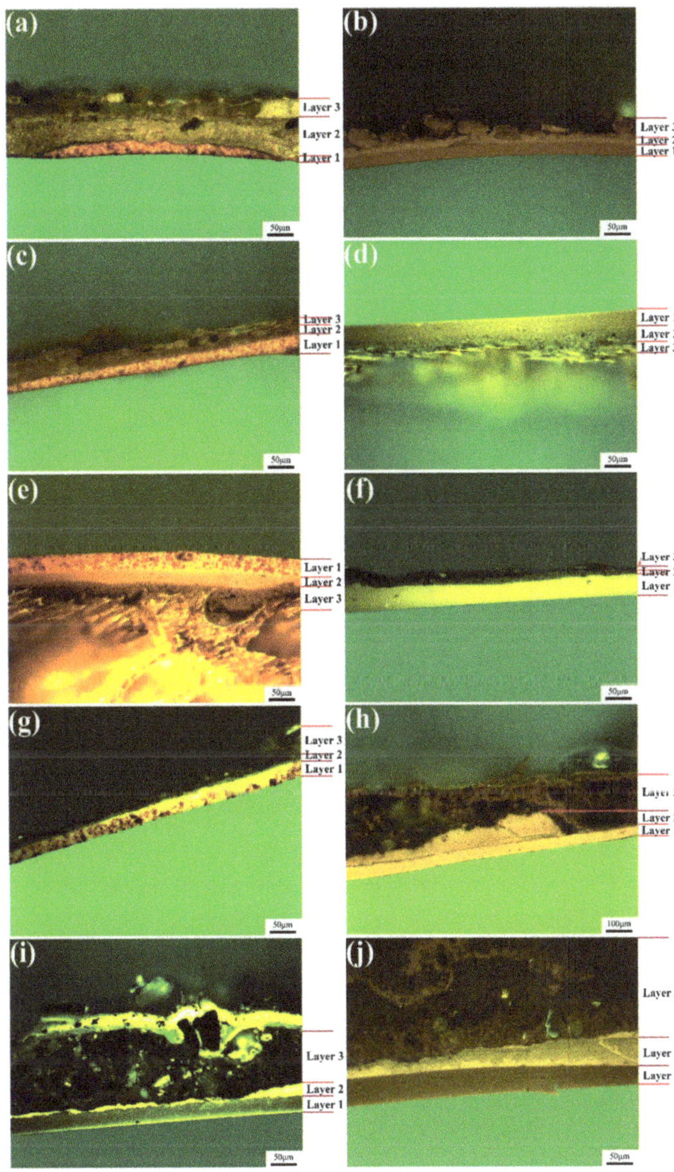

Figure 2. Cross-sectional photos of (**a**) Sample 1, (**b**) Sample 2 (black), (**c**) Sample 2 (red), (**d**) Sample 3 (black), (**e**) Sample 3 (red), (**f**) Sample 4 (black), (**g**) Sample 4 (red), (**h**) Sample 5, (**i**) Sample 6, and (**j**) Sample 7 under blue light.

As shown in Figure 2, the lacquering techniques of lacquer Lian underwent minimal changes during nearly 1800 years from the Warring States Period to the Yuan Dynasty. Firstly, the ground layer was painted on the lacquer bodies to fill the pores. Secondly, the undercoat layer was used to cover the defects of the ground layer. Lastly, a colored layer or a lacquer layer was used as the surface finish layer to decorate the lacquerware. However, the thickness of the same layer in different eras varied greatly; for instance, the red colored layer became thicker from the Warring States Period to the Han Dynasty, as discussed above.

3.2. Pigment Analysis

According to Xiushilu, cinnabar, red ochre, and crimson melanterite could be used to make red lacquer. Raman analysis was used to clarify the minerals in the red pigments. Figure 3 shows the Raman spectra of the red pigments. All the four red pigments show similar Raman spectra, with peaks at about 254 and 343 cm^{-1}, which are characteristic bands of cinnabar. Due to its excellent color, gloss, and anti-corrosion properties, cinnabar is the best choice for making red lacquer. The results were consistent with previous studies [12,23].

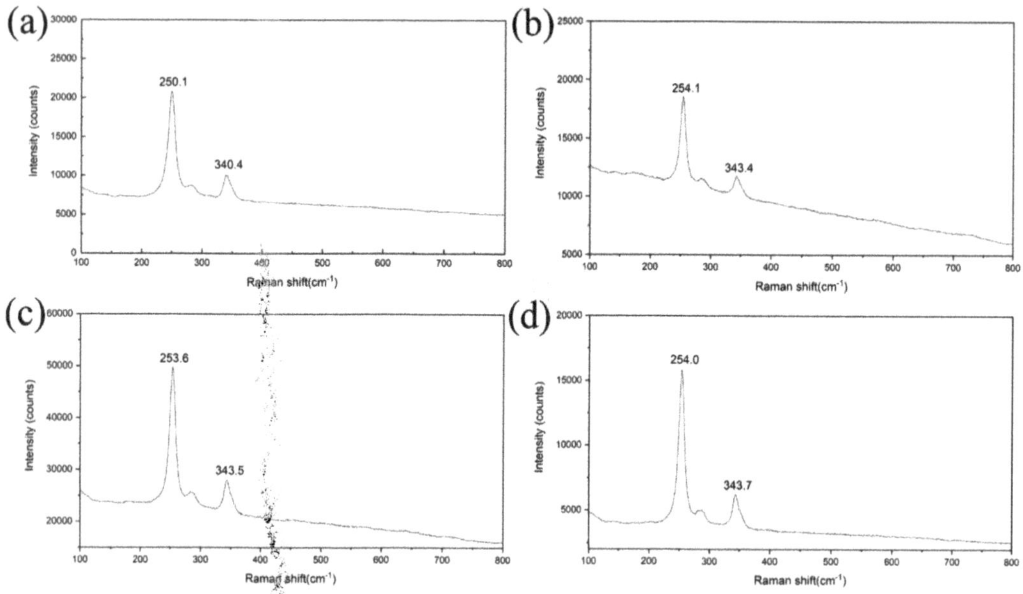

Figure 3. Raman spectra of red pigments of (**a**) Sample 1, (**b**) Sample 2, (**c**) Sample 3, and (**d**) Sample 4.

3.3. Lacquer Film Analysis

In order to identify the types of lacquer and oil in different layers of the seven samples, THM-Py-GC/MS analysis was carried out. All the seven samples showed similar chromatographic profiles; only the chromatographic profile of the red layer of sample 1 is presented in Figure 4, and the rest are shown in Supplementary Materials. The presence of lacquer was assessed by using EICs, as shown in Figure 4. Characteristic pyrolysis products were confirmed. Aliphatic hydrocarbons (C, m/z 55 and 57, Figure 4b,c) are present from 1-Decene (C9:1) to 1-Pentadecene (C15:1), peaking at 1-Tetradecene (C14:1), and present from decane (C10) to pentadecane (C15), peaking at pentadecane (C15); the alkylbenzenes (B, m/z 91, Figure 4d) show a decreasing profile from benzene propyl- (B3) to benzene octyl- (B8). These results are identical to those of urushi [24]. In addition, 1,2-Dimethoxy-3-pentadec-8-enylbenzene (P1, 3-pentadecenyl-catechol) and 1,2-Dimethoxy-3-pentadecylbenzene (P2, 3-pentadecyl-catechol) were detected in the red layer; both are characteristic components of urushi [25,26]. Based on the above results, it could be concluded that the lacquer of the seven samples was Chinese lacquer (urushi).

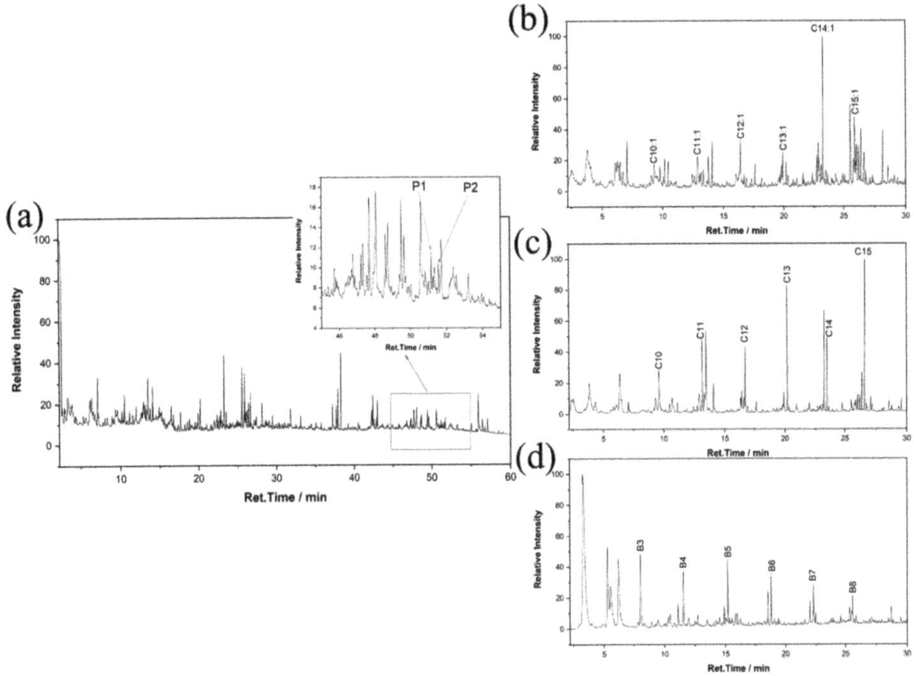

Figure 4. Chromatographic profiles obtained using THM-Py–GC/MS of red layer of sample 1: (**a**) total ion pyrogram; (**b**) m/z 55 extracted ion pyrogram; (**c**) m/z 57 extracted ion pyrogram; (**d**) m/z 91 extracted ion pyrogram.

Figure 5 shows the relative concentration of fatty acids detected from the red layers. It can be seen that a large number of mono-carboxylic acids and di-carboxylic acids were detected, especially nonanedioic acid (C8) and octanedioic acid (C9), which are typical pyrolytical and aging products of dying oil. These results indicate that drying oil was used in the red layer at least during the Warring States Period, which is identical to previous reports [19]. Lacquer could be easily mixed with pigments by drying oil to obtain colorful lacquerware.

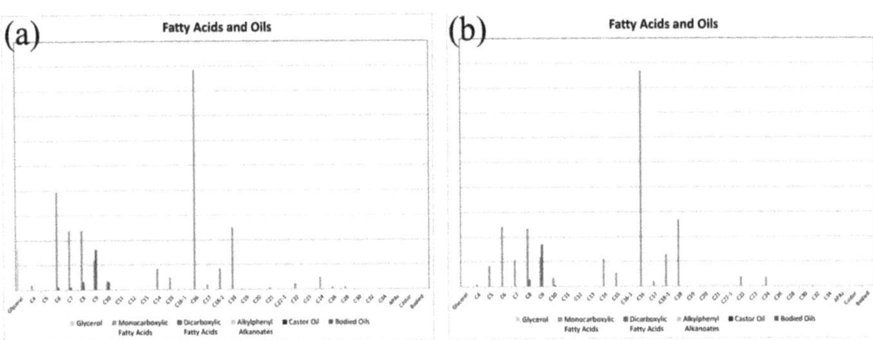

Figure 5. *Cont.*

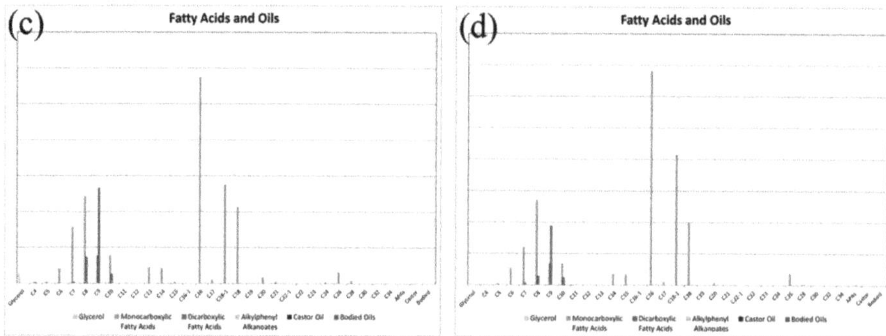

Figure 5. The relative concentration of fatty acids in red layers of (**a**) Sample 1, (**b**) Sample 2, (**c**) Sample 3, and (**d**) Sample 4.

Figure 6 shows the relative concentration of fatty acids detected from black layers. There is no drying oil in the black layers of lacquer Lian from the Warring States Period, as shown in Figure 6a,b. Castor oil, which is one kind of drying oil, was detected in the black layers of lacquerwares from the Han Dynasty and Song Dynasty, as shown in Figure 6c,d. Figure 6e,f indicate that there is drying oil in the black layers of lacquerwares from the Song Dynasty and Yuan Dynasty. Drying oil can decelerate the hardening process of lacquer while enhancing the luster and flexibility of the lacquer, thus elevating the quality of lacquerware.

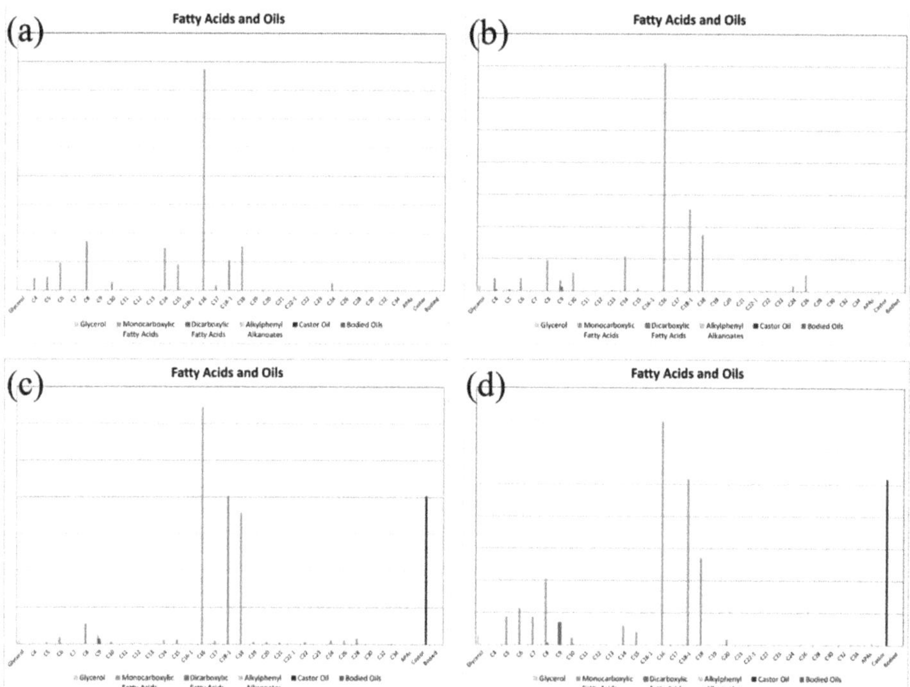

Figure 6. *Cont.*

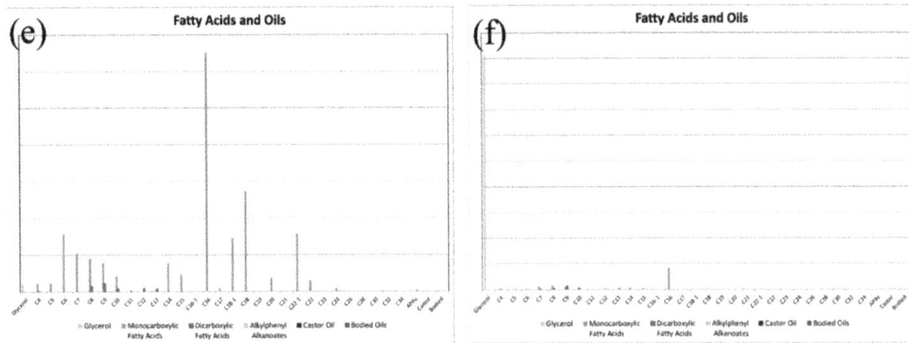

Figure 6. The relative concentration of fatty acids in black layers of (**a**) Sample 2, (**b**) Sample 3, (**c**) Sample 4, (**d**) Sample 5, (**e**) Sample 6, and (**f**) Sample 7.

The relative concentration of fatty acids detected from the ground layers is shown in Figure 7. Drying oil was used in the ground layer for Sample 1, but not for Sample 2, as shown in Figure 7a,b. Figure 7c,d shows that drying oil was widely used in ground layers in the Han Dynasty. Castor oil was detected in the ground layer of Sample 5 (Figure 7e). Since methyl alkylphenyl alkanoate (APA) is the marker component of boiled tung oil, it could be concluded that boiled tung oil was used in the ground layers of Samples 6 and 7, based on the results of Figure 7f,g. Boiled tung oil, serving as an organic binder for ground layers, emerged in the Song Dynasty at the latest, in accordance with other findings [27].

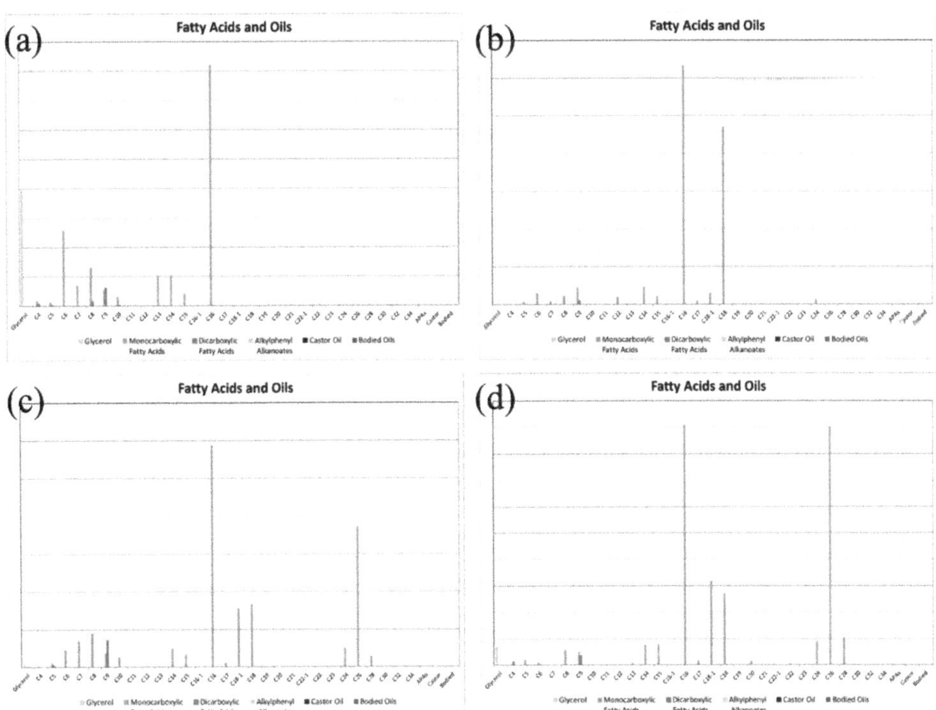

Figure 7. *Cont.*

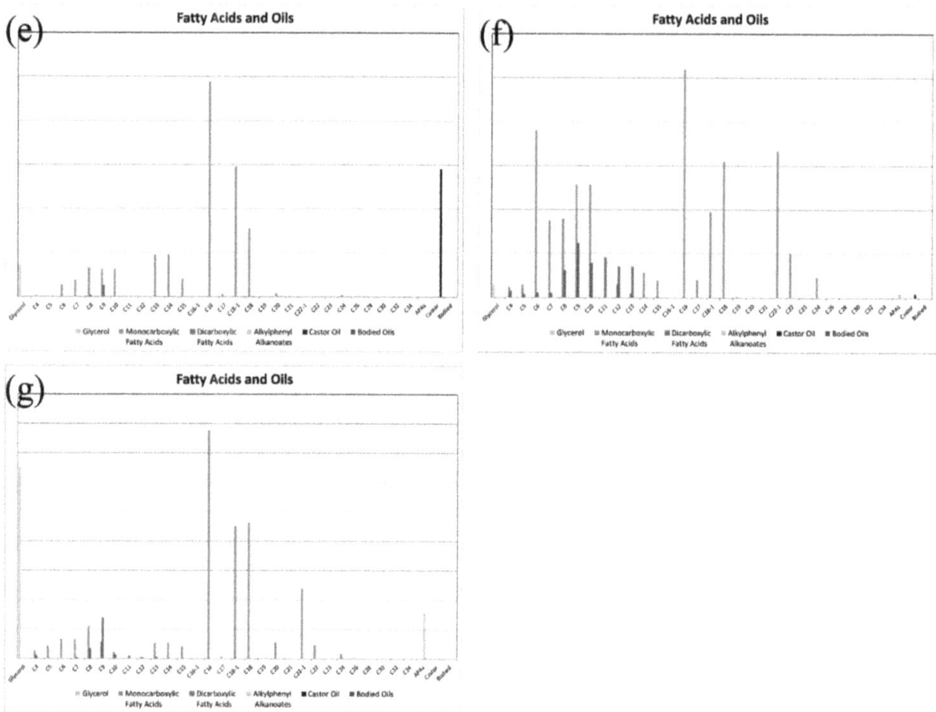

Figure 7. The relative concentration of fatty acids in ground layers of (**a**) Sample 1, (**b**) Sample 2, (**c**) Sample3, (**d**) Sample 4, (**e**) Sample 5, (**f**) Sample 6, and (**g**) Sample 7.

4. Conclusions

Seven lacquer Lian samples from the Warring States Period, Han Dynasty, Song Dynasty, and Yuan Dynasty were comprehensively investigated using a scientific analytical approach. The cross-section photos indicated that the lacquer Lian included a pigment layer, undercoat layer, and ground layer, and lacquering techniques underwent minimal changes from the Warring States Period to the Yuan Dynasty, which indicates the inheritance of lacquer techniques. Raman analysis clarified that the red pigment was cinnabar. The THM-Py-GC/MS results indicated that all the lacquerware items were coated with lacquer sap collected from a Rhus vernicifera lacquer tree. Drying oil was mixed with lacquer sap to increase the luster and elastic of lacquer film. Boiled tung oil was found in the ground layer of lacquerware from the Song Dynasty and Yuan Dynasty, reflecting the development of lacquer materials. This study not only provides a better understanding of the inheritance and evolution of Chinese lacquering techniques and lacquering materials, but also provides scientific support for the preservation and conservation of unearthed lacquerware.

Supplementary Materials: The following supporting information can be downloaded at: https://www.mdpi.com/article/10.3390/coatings13101750/s1.

Author Contributions: H.W., the lead author, carried out the Raman experiments on the samples, analyzed the data, and designed the article. The second author, Y.Z., carried out THM-Py-GC/MS of the samples. The third author, B.F., provided archaeological samples and their archaeological information. The corresponding author and the fourth author, J.D., designed the article and funded the research. All authors have read and agreed to the published version of the manuscript.

Funding: This work was financially supported by the National Key R&D Program of China [No. 2019YFC1520300].

Institutional Review Board Statement: Not applicable.

Informed Consent Statement: Not applicable.

Data Availability Statement: Not applicable.

Conflicts of Interest: The authors declare that they have no known competing financial interests or personal relationships that could have appeared to influence the work reported in this paper.

References

1. Niimura, N.; Miyakoshi, T.; Onodera, J.; Higuchi, T. Characterization of Rhus vernicifera and Rhus succedanea lacquer films and their pyrolysis mechanisms studied using two-stage pyrolysis-gas chromatography/mass spectrometry. *J. Anal. Appl. Pyrolysis* **1996**, *37*, 199–209. [CrossRef]
2. Niimura, N.; Miyakoshi, T.; Onoderra, J.; Shiguchi, T. Structural studies of Melanorrhoea usitate lacquer film using two-stage pyrolysis/gas chromatography/mass spectrometry. *Rapid Commun. Mass Spectrom.* **1996**, *10*, 1719–1724. [CrossRef]
3. Tamburini, D.; Bonaduce, I.; Colombini, M.P. Characterization and identification of urushi using in situ pyrolysis/silylation–gas chromatography–mass spectrometry. *J. Anal. Appl. Pyrolysis* **2015**, *111*, 33–40. [CrossRef]
4. Kumanotani, J. Urushi (oriental lacquer)—A natural aesthetic durable and future-promising coating. *Prog. Org. Coat.* **1995**, *26*, 163–195. [CrossRef]
5. Lu, R.; Kamiya, Y.; Miyakoshi, T. Applied analysis of lacquer films based on pyrolysis-gas chromatography/mass spectrometry. *Talanta* **2006**, *70*, 370–376. [CrossRef]
6. Sung, M.; Jung, J.; Lu, R.; Miyakoshi, T. Study of historical Chinese lacquer culture and technology—Analysis of Chinese Qin-Han dynasty lacquerware. *J. Cult. Herit.* **2016**, *21*, 889–893. [CrossRef]
7. Wu, M.; Zhang, B.; Jiang, L.; Wu, J.; Sun, G. Natural lacquer was used as a coating and an adhesive 8000 years ago, by early humans at Kuahuqiao, determined by ELISA. *J. Archaeol. Sci.* **2018**, *100*, 80–87. [CrossRef]
8. Wu, M.; Zhang, Y.; Zhang, B.; Li, L. Study of colored lacquerwares from Zenghou Yi Tomb in Early Warring States. *New J. Chem.* **2021**, *45*, 9434–9442. [CrossRef]
9. Zhao, Y.; Tang, Y.; Tong, T.; Sun, Z.; Yu, Z.; Zhu, Y.; Tong, H. Red lead degradation: Monitoring of color change over time. *New J. Chem.* **2016**, *40*, 3686–3692. [CrossRef]
10. Watson, A.Y.; Valberg, P.A. Carbon black and soot: Two different substances. *AIHAJ* **2001**, *62*, 218–228. [CrossRef]
11. Zheng, L.; Wang, L.; Zhao, X.; Yang, J.; Zhang, M.; Wang, Y. Characterization of the materials and techniques of a birthday inscribed lacquer plaque of the qing dynasty. *Herit. Sci.* **2020**, *8*, 116. [CrossRef]
12. Ma, X.; Shi, Y.; Khanjian, H.; Schilling, M.; Li, M.; Fang, H.; Cui, D.; Kakoulli, I. Characterization of Early Imperial Lacquerware From The Luozhuang Han Tomb, China. *Archaeometry* **2017**, *59*, 121–132. [CrossRef]
13. De Oliveira, L.F.C.; Edwards, H.G.M.; Frost, R.L.; Kloprogge, J.T.; Middleton, P.S. Caput mortuum: Spectroscopic and structural studies of an ancient pigment. *Analyst* **2002**, *127*, 536–541. [CrossRef] [PubMed]
14. Karpova, E.; Nefedov, A.; Mamatyuk, V.; Polosmak, N.; Kundo, L. Multi-analytical approach (SEM-EDS, FTIR, Py-GC/MS) to characterize the lacquer objects from Xiongnu burial complex (Noin-Ula, Mongolia). *Microchem. J.* **2017**, *130*, 336–344. [CrossRef]
15. Wang, X.; Zhen, G.; Hao, X.; Zhou, P.; Wang, Z.; Jia, J.; Gao, Y.; Dong, S.; Tong, H. Micro-Raman, XRD and THM-Py-GC/MS analysis to characterize the materials used in the Eleven-Faced Guanyin of the Du Le Temple of the Liao Dynasty, China. *Microchem. J.* **2021**, *171*, 106828. [CrossRef]
16. Tamburini, D.; Sardi, D.; Spepi, A.; Duce, C.; Tinè, M.R.; Colombini, M.P.; Bonaduce, I. An investigation into the curing of urushi and tung oil films by thermoanalytical and mass spectrometric techniques. *Polym. Degrad. Stab.* **2016**, *134*, 251–264. [CrossRef]
17. Honda, T.; Lu, R.; Sakai, R.; Ishimura, T.; Miyakoshi, T. Characterization and comparison of Asian lacquer saps. *Prog. Org. Coat.* **2008**, *61*, 68–75. [CrossRef]
18. Tamburini, D.; Pescitelli, G.; Colombini, M.P.; Bonaduce, I. The degradation of Burmese lacquer (thitsi) as observed in samples from two cultural artefacts. *J. Anal. Appl. Pyrolysis* **2017**, *124*, 51–62. [CrossRef]
19. Fu, Y.; Chen, Z.; Zhou, S.; Wei, S. Comparative study of the materials and lacquering techniques of the lacquer objects from Warring States Period China. *J. Archaeol. Sci.* **2020**, *114*, 105060. [CrossRef]
20. Yuasa, K.; Honda, T.; Lu, R.; Hachiya, T.; Miyakoshi, T. Analysis of Japanese ancient lacquerwares excavated from Jōmon period ruins. *J. Anal. Appl. Pyrolysis* **2015**, *113*, 73–77. [CrossRef]
21. Wei, S.; Pintus, V.; Pitthard, V.; Schreiner, M.; Song, G. Analytical characterization of lacquer objects excavated from a Chu tomb in China. *J. Archaeol. Sci.* **2011**, *38*, 2667–2674. [CrossRef]
22. Wang, X.; Hao, X.; Zhao, Y.; Tong, T.; Wu, H.; Ma, L.; Shen, X.; Tong, H. Systematic study of the material, structure and lacquering techniques of lacquered wooden coffins from the Eastern Regius Tombs of the Qing Dynasty, China. *Microchem. J.* **2021**, *168*, 106369. [CrossRef]
23. Jin, P.J.; Hu, Y.L.; Ke, Z.B. Characterization of lacquer films from the middle and late Chinese warring states period 476-221BC. *Microsc. Res. Tech.* **2017**, *80*, 1344–1350. [CrossRef] [PubMed]
24. Tamburini, D.; Bonaduce, I.; Ribechini, E.; Gallego, C.; Perez-Arantegui, J. Challenges in the data analysis of Asian lacquers from museum objects by pyrolysis gas chromatography/mass spectrometry. *J. Anal. Appl. Pyrolysis* **2020**, *151*, 104905. [CrossRef]

25. Niimura, N. Determination of the type of lacquer on East Asian lacquer ware. *Int. J. Mass Spectrom.* **2009**, *284*, 93–97. [CrossRef]
26. Schilling, M.R.; Heginbotham, A.; van Keulen, H.; Szelewski, M. Beyond the basics: A systematic approach for comprehensive analysis of organic materials in Asian lacquers. *Stud. Conserv.* **2016**, *61*, 3–27. [CrossRef]
27. Hao, X.; Schilling, M.R.; Wang, X.; Khanjian, H.; Heginbotham, A.; Han, J.; Auffret, S.; Wu, X.; Fang, B.; Tong, H. Use of THM-PY-GC/MS technique to characterize complex, multilayered Chinese lacquer. *J. Anal. Appl. Pyrolysis* **2019**, *140*, 339–348. [CrossRef]

MDPI
St. Alban-Anlage 66
4052 Basel
Switzerland
www.mdpi.com

Coatings Editorial Office
E-mail: coatings@mdpi.com
www.mdpi.com/journal/coatings

www.ingramcontent.com/pod-product-compliance
Lightning Source LLC
Chambersburg PA
CBHW062038220926
51850CB00010B/2596
* 9 7 8 3 7 2 5 8 0 4 2 9 0 *